PENGUIN BOOKS

Me, Myself, and Why

JENNIFER OUELLETTE is a science journalist and the author of three previous books, *The Calculus Diaries*, *The Physics of the Buffyverse*, and *Black Bodies and Quantum Cats*. Her work has appeared in *The Washington Post*, *Discover*, *Slate*, *Salon*, and *Nature*, among other publications. She writes a science and culture blog called *Cocktail Party Physics* on the Web site of *Scientific American*. Ouellette served from 2008 to 2010 as the director of the Science and Entertainment Exchange, a program of the National Academy of Sciences that aims to foster creative collaborations between scientists and entertainment-industry professionals. She has also been the Journalist in Residence at the Kavli Institute of Theoretical Physics in Santa Barbara and an instructor at the Santa Fe Science Writing Workshop. Ouellette holds a black belt in jujitsu and lives in Los Angeles with her husband, Caltech physicist Sean Carroll.

Praise for Jennifer Ouellette

The Calculus Diaries

"Ouellette makes math palatable with a mix of humor, anecdote, and enticing facts. . . . Using everyday examples, such as petrol mileage and fairground rides, she makes even complex ideas such as calculus and probability appealing." —*Nature*

"This dash through a daunting discipline bursts with wry wit. Ouellette uses differential equations to model the spread of zombies and derivatives to craft the perfect diet. Sassy throughout, she reserves special barbs for subprime mortgage holders: 'Chances are they weren't doing the math.' " —*Discover*

"A great primer for anyone who needs to get over their heebie-jeebies about an upcoming calculus class, or for anyone who's ever wondered how calculus fits into everyday life and wants to be entertained, too!"
—Danica McKellar, *New York Times* bestselling author of *Math Doesn't Suck* and *Hot X: Algebra Exposed*

"I haven't had this much fun learning math since I watched the Count on *Sesame Street* when I was three. And the Count never talked about log flumes or zombies. So *The Calculus Diaries* wins the day."
—A. J. Jacobs, *New York Times* bestselling author of *The Know-It-All* and *Drop Dead Healthy*

"Ouellette is every English major's dream math teacher: funny, smart, infected with communicable enthusiasm—and she can rock a Buffy reference. In this book, she hastens the day when more people are familiar with an integral function than with Justin Bieber."
—Peter Sagal, host, NPR's *Wait, Wait Don't Tell Me*, and author of *The Book of Vice*

"Wonderful and compulsively readable . . . Ouellette finds the signature of mathematics—and especially calculus, of course—in the most unexpected places, from the gorgeously lunatic architecture of Spain's Antonio Gaudi to the shimmering arc of waves on a beach. . . . Her ever clear and always stimulating voice is a perfect match to the subject. *The Calculus Diaries* is a tour de force."

—Deborah Blum, author of *The Poisoner's Handbook*

"As amusing as it is enlightening . . . Ouellette steers us so gently we think we're gliding along on our own."

—Michael Sims, author of *Adam's Navel*

"If you ever thought that math was useless, read this book. Want to survive a zombie attack? Win at craps? Beat a zombie at craps? Well, listen to Jennifer Ouellette. The math she describes might just be your best hope if you don't want your brains to be gobbled by the undead."

—Charles Seife, author of *Zero: Biography of a Dangerous Idea*

"A charming and gentle introduction to important mathematical concepts and their relevance to everyday life."

—Leonard Mlodinow, author of *The Drunkard's Walk: How Randomness Rules Our Lives*

The Physics of the Buffyverse

"Blending fizzy pop culture with serious science . . . Ouellette makes an earnest effort to introduce the laws of physics to couch potatoes in a relatively painless way."

—*The New York Times Book Review*

"If you dig science, vampires and the like, give *The Physics of the Buffyverse* a try." —*USA Today*

"Ouellette presents a strong case for many of the seemingly impossible aspects of the world Buffy and her friends inhabit. All the while, she makes the science accessible, guaranteeing that fans of the show will be receptive." —*Booklist*

Black Bodies and Quantum Cats

"Bursts with answers for curious adults . . . Employing contemporary cultural icons like the movie *Addams Family Values* and *The Da Vinci Code*, Ouellette explains the principles behind acceleration and ancient geometrical anomalies. . . . Ouellette shines when she pulls analogies from real life to explain, for example, why blackouts are more likely since the deregulation of the power industry, in prose that is engaging and economical."
—*The Washington Post*

"Readers of these pieces will feel Ouellette's companionship as a fellow layperson sharing her interest in physics history. Hooking the audience with some movie or science-fiction novel . . . her entertaining explications encourage generalists to give physics a try."
—*Booklist*

"Remarkably fresh and immensely readable. . . . All major theories and breakthroughs, along with the personalities that brought them to life (including a particularly ruthless Thomas Edison and a resourceful patent clerk named Chester Carlson, who built the first photocopier in his Astoria, New York, kitchen), are presented clearly by the reader's pop-culture escort. It is a credit to Ouellette that, as the reader progresses into more complex theories, the TV and movie references aren't nearly as interesting as the science."
—*Publishers Weekly*

Me, Myself, and Why

SEARCHING FOR
THE SCIENCE OF SELF

Jennifer Ouellette

PENGUIN BOOKS

PENGUIN BOOKS

Published by the Penguin Group
Penguin Group (USA) LLC
375 Hudson Street
New York, New York 10014

USA | Canada | UK | Ireland | Australia | New Zealand | India | South Africa | China
penguin.com
A Penguin Random House Company

First published in Penguin Books 2014

LIBRARY OF CONGRESS CATALOGING-IN-PUBLICATION DATA
Ouellette, Jennifer.
Me, myself, and why : searching for the science of self / Jennifer Ouellette.
pages   cm
Includes bibliographical references and index.
ISBN 978-0-14-312165-7
1. Self psychology.   2. Self-actualization (Psychology)
3. Identity (Philosophical concept)   I. Title.
BF697.Q778   2014
155.2—dc23   2013034514

Printed in the United States of America
1   3   5   7   9   10   8   6   4   2

Set in Sabon MT Std
Designed by Elke Sigal

For my parents, Paul and Jeanne

One may understand the cosmos, but never the ego; the self is more distant than any star.

—G. K. CHESTERTON,
"THE LOGIC OF ELFLAND," *Orthodoxy* (1908)

CONTENTS

ACKNOWLEDGMENTS

Several years ago, I resolved to confront my lifelong math phobia head-on and recounted that experience in *The Calculus Diaries*. That's how I discovered that the story I'd been telling myself—that I was bad at math—wasn't based on fact. When I went back and checked the records, I discovered that I earned A's in all my math classes. It got me thinking about everything that goes into how we define ourselves and craft a personal narrative and how this in turn influences not just our identity but the choices we make and what we think we can and cannot do. Ultimately this led to a book project exploring the science of the self.

I soon realized this was an impossibly broad subject. Massive tomes and stacks of scholarly papers have been written about each of the topics covered herein, involving decades, if not centuries, of research. Boiling all that material down into a concise, reader-friendly overview was the very definition of a Herculean task. The end result is a highly selective, often personal account that I hope will give general readers the scientific lay of the land and provide some fodder for thoughtful reflection on our most deeply ingrained assumptions about ourselves.

The number of scientists who generously shared their expertise is downright humbling. I could not have written this book without their patient input. Ulrike Heberlein gave me my first glimpse of drunken fruit flies under a microscope. David Poeppel spent an afternoon discussing the neurological intricacies of the self and let me observe an MEG scan in the bargain. Patricia Churchland welcomed me into her home and enlightened me on the finer philosophical points of consciousness. Anne Wojcicki, Brian Naughton, Joanna Mountain, and Catherine Afarian of 23andMe took time out of their busy schedules to give me an overview of their venture. I still log in to the site from time to time and remain a fan of the service.

I had a lively, entertaining visit with David Eagleman, who provided a brain scan and allowed me to chat with his entire lab over a Texas-sized noodle bowl: Mingbo Cai, Josh Jackson, Sean Judge, Scott Novich, Ricky Savjani, Steffie Tomson, and Don Vaughn. Kathleen McDermott and Henry Roediger introduced me to the entire psychology department at Washington University in St. Louis, many of whom agreed to be interviewed, thereby giving me a crash course in the field: Katie Arnold, Erika Carlson, Bridgid Finn, Adrian Gilmore, Joshua Jackson, Randy Larsen, Lori Markson, Tom Oltmanns, Mike Strube, and Simine Vazire. I spent an afternoon with Jackie Morie of the University of Southern California enthusing over avatars and identity in virtual worlds. Ricardo Gil da Costa spent two hours going over brain scans and the neural correlates of the self with me, and Uri Hasson patiently walked me through his work on neural coupling in a La Jolla diner, sketching helpful diagrams on the paper placemat.

Thanks also to the following for granting interviews: Jeremy Bailenson, Michael Bailey, Sven Bocklandt, Matthew Botvinick, Robin Carhart-Harris, Laura Case, Anthony Chemero,

Meredith Chivers, Danielle Dick, Rick Doblin, David Feather-stone, Andrew Gerber, Rob Goldstone, Sam Gosling, Nicholas Grahame, Carla Green, John Halpern, Bernhard Hommel, Christof Koch, Joseph LeDoux, Dan McAdams, David Nichols, Cibele Ruas, Juan Sanchez-Ramos, Sebastian Seung, Lindsay Squeglia, and Bradley Voytek.

Extra-special thanks to the pseudonymous "Rory," for shar-ing his personal transition story-in-progress; and to Meg Bow-ers and Petra Boynton for helping me navigate the treacherous waters of gender identity and sexual orientation.

One never knows when a casual conversation will turn out to lead one in fruitful directions, so I must also thank Charlie Jane Anders, Misha Angrist, Allyson Beatrice, Deborah Blum, Bethany Brookshire, Gay Crooks, George Djorgovski, David Dobbs, Raissa D'Souza, Shari Steelsmith-Duffin, Esther Dy-son, A. V. Flox, Jason Goldman, Kieran Healy, JoAnne Hewett, Alan Klein, Maria Konnikova, John de Lancie, Joshua Landy, Tom Levenson, Michael Lill, Ben Lillie, Robin Lloyd, Peri Lyons, Malcolm MacIver, Jennifer McCreight, Usha McFar-ling, Bob Mondello, George Musser, Jr., Annalee Newitz, An-nie Murphy Paul, Laurie Paul, Nick and Susan Pritzker, John Rennie, Sherry Reson, Cassie Rodenberg, Michael Russell, Nick Sagan, Cara Santa-Maria, John Scalzi, Carlos Schroeder, Steve Silberman, Michael Sims, Brian Switek, Melanie Tannen-baum, Carol Tavris, Holly Tucker, David Wallace, Nick War-ner, Carolee Winstein, Ed Yong, Robin Yuan, Amos Zeeburg, Philip Zimbardo, and Carl Zimmer.

This is my fourth book for Penguin, in what has proved to be a fruitful partnership. I am grateful to my editor, Chris Russell, for providing focus to my meandering prose as I strug-gled to pull together so many disparate threads. As always, I owe many thanks to my agent, Mildred Marmur, who looks out for my best interests.

My sister, Amy Ouellette English, doggedly transcribed many hours of interviews, with Jacqueline Smay picking up the slack when needed. Lee Kottner reviewed the draft manuscript and gave her usual excellent pointers. My husband, Sean Carroll, gently pointed out logical inconsistencies and consoled me during those dark nights of the authorial soul, when I was convinced I could never do the topic justice. Every day he makes my life that much richer for being part of it. Love and gratitude go to my parents, Paul and Jeanne Ouellette, whose support has never wavered. Mom, Dad, this one's for you.

Finally, there is that unknown woman who endured all the social pressures and emotional anguish that inevitably accompany giving up a child for adoption. I hope she would look at the woman I became and feel that her sacrifice was not in vain.

Me, Myself, and Why

Prologue

Since I was five, I've known that I was adopted, which is a politically correct term for being clueless about one's own origins.

—JODI PICOULT, *Handle with Care*

When I was five years old, I held my baby sister for the very first time—not in the sterile environs of a hospital maternity ward, but in the slightly musty waiting room of a Midwestern adoption agency, when she was already several weeks old. I knew that I, too, was adopted, as was my older brother, even if I didn't fully grasp exactly what that meant. Adoption was simply one more fact of life. It certainly postponed all those awkward questions about where babies come from, leaving no need for the invocation of a mythical stork, since for years I assumed that everyone just went to the local adoption store to pick out a new baby, much like buying groceries at the supermarket.

My biological origins were little more than a point of mild curiosity until I moved to New York City after college and discovered that other people were far less blasé about the subject. Perhaps it struck them as exotic. A few held strong opinions on how I ought to feel about being adopted, often expressing mild shock at my apparent lack of interest: "But don't you want to know?" they would gasp, wide-eyed. "Not really," I'd shrug in reply and try to change the subject.

I never had the urge to embark on the long, obstacle-laden process of seeking out my biological parents. I bristled at the mere suggestion that this wasn't my "real" family (and still do). My mother once confessed that for the first two years of my life, her greatest fear was that my biological mother would show up, un-announced, and take me away from her. I saw no need to dredge up those old emotions. Moreover, my natural reserve made me balk at the prospect of ferreting out a perfect stranger—one who, for all I know, had spent more than four decades trying to forget a painful chapter in her youth—simply out of idle curiosity. I ex-perienced no psychologically crippling abandonment issues, nor had I felt gaping emotional holes in my life or confronted pressing medical issues—nothing to warrant the intrusion. I knew very well who I was; it never occurred to me to wonder how I came to be that person in the first place.

Then one day, my mother called and told me she had just mailed me all the documentation she had on the details of my birth: "You're all grown up now, and it's high time I handed them on to you." She hesitated a moment, then added, "I just want to warn you that the adoption papers—well, they show your birth name."

Another name? My interest piqued. A few days later, I opened a plain manila envelope to find a faded birth certificate, duly notarized, and a thin sheet of onionskin paper with neatly typed instructions on my care and feeding ("This baby eats ce-reals twice a day and prefers to sleep on her stomach"). The sheet included some scant details about my biological parents.

She was in her late teens; of French, Irish, and German descent; played the violin; and graduated as co-salutatorian of her high school class. He was in his early twenties, Norwegian, and athletic.

Finally, there was the mimeographed decree of adoption, legally changing my name from the one my birth mother gave me to the one I bear today. I admit to being momentarily nonplussed by the sight of that unfamiliar name and surname. But the feeling quickly passed, because I didn't feel as if the "petitioner" named in those documents ("a minor") was really me—not anymore. William James defined the self as "the sum total of all that a man can call his, not only his body and his psychic powers, but his clothes, and his wife and children, his ancestors and friends, his reputation and works, his lands and horses, and yacht and bank account."* Our given names are simply what we call the totality of everything that makes us who we are, and that sum total is constantly in flux.

Still, names possess tremendous psychological power, and in that sense, the manila envelope contained the fragments of a self that might have been. For the first time, I started pondering in earnest the age-old questions about nature and nurture. How much of who we become is due to genetic factors, and how much is shaped by our experiences and environment—and ultimately, by our choices? Would we be markedly different if we had different parents and formative experiences? Or would we be largely the same? I'm sure my biological mother also wondered about the road not taken. How did giving away her child in her late teens mold and shape her for the rest of her life? How might she have been different, had she made a different choice?

---

* James moved in privileged circles.

There is a reason that alternate realities populated by doppel-gängers are among the most common tropes in science fiction and fantasy.

Curiosity about my own life prompted my interest in the science behind the self—namely, how a fully conscious, unique individual emerges from the genetic primordial ooze. It is a daunting subject. Each of us has *a* self, and because of this we may think we understand *the* self. We don't. We can't even agree on what we mean by "the self." There is no common vocabulary, making it difficult to discuss the science in a way that allows for fine demarcations of meaning across very different disciplines.

Ask a physicist, and she might tell you that you are a collection of atoms (made up of even smaller subatomic particles) obeying well-defined laws as they interact via forces. Ask a biologist, and he might tell you that you are an organism whose form is the result of complicated processes involving genes and proteins and countless biochemicals, working in tandem with environmental factors to shape a unique individual. A neuroscientist might point to the intricate wiring of the brain as the essence of self, while a social psychologist might say we define who we are via our interactions with others and our place in society—or by our "stuff," the material objects with which we surround ourselves. And if you ask certain French philosophers, you might be told that the self is primarily a cultural construct, and reality itself is just a persistent illusion, at which point your head will explode.

At least one thing is clear: the nature-versus-nurture dichotomy is simplistic and outdated, long ago confined by serious scientists to the dustbins of history. As Francis Galton framed the question back in 1874, "Nature is all that a man brings with himself into the world; nurture is every influence that affects him after his birth." Galton was one of the first to conduct a formal study of identical twins, now a mainstay of such research. Yet even he admitted that this definition is "a convenient jingle of words, for it separates under two distinct heads the innumerable elements of which personality is composed."

There is no question that "blood will out" for certain traits: I am within an inch of height and roughly the same weight as my biological mother before her pregnancy, with similar skin tone and hair color. Surely it is no coincidence that I was also bright and musically inclined, latching onto piano, guitar, and voice lessons with far more enthusiasm than my adoptive siblings. I blame my strapping Norwegian biological father for my broad shoulders and back, although his athleticism may well be reflected in my love for swimming, bicycling, and martial arts. Once, while attending a conference in Copenhagen, I stood next to a Norwegian woman in the restroom as we washed our hands. Our eyes met briefly in the mirror, and I was startled by the marked similarities in our facial features. (Adopted children are unaccustomed to strong familial resemblance.)

But there are also less tangible traits that undeniably reflect, to varying degrees, the influence of my early home environment and unique life experiences. When I developed blisters on vacation while walking all over Florence and Paris, I refused to stop walking every day, even when a mild infection set in. "Why are you doing this to yourself?" my baffled husband asked as I hobbled down the street. "I didn't come all the way to Europe to take cabs everywhere or stay in the hotel room!" I retorted. It is certainly plausible that this stubbornness has a genetic basis, but my sister visited us in Los Angeles a few weeks later with a badly twisted ankle and insisted on traipsing around Six Flags

Magic Mountain anyway, for exactly the same reason. Since my sister and I are both adopted, and genetically unrelated, this may be a shared trait, the result of being raised in a family that encouraged stoicism in the face of adversity. (The Ouellette family motto, spoken through gritted teeth, is "We're Fine.") More likely, it is a combination of the two.

It's nature *and* nurture—not one or the other—locked in an intricate dance, whether we are talking about eye color, food preferences, disease risk factors, personality traits, alcohol tolerance, behavior, gender identity and expression, or sexual orientation. While genes are deterministic up to a point, they are far from destiny. (Except, I learned, when it comes to earwax consistency. Your earwax is your destiny.) Genes interact both with other genes and with myriad environmental factors, including parenting, peer pressure, cultural influences, unique life experiences, and even the hormones that contribute to our development while still in the womb. The debate now centers on matters of degree: how much of the observed variation among human beings for a given trait is due to each influencing factor.

Genes form the biological scaffolding upon which all else is constructed, most notably the brain. While we are still in the womb, genes are churning out proteins to regulate body and brain development, and this, in turn, influences any number of traits. That is nature's contribution. Nurture also plays a vital role, primarily through our experiences with the world. New synaptic connections are constantly being formed over the course of our lives, most of which are unique to us as individuals, further distinguishing one person from another—even in the case of identical twins. "Nature and nurture are really two ways of doing the same thing: wiring up synapses," neuroscientist Joseph LeDoux observed in *The Synaptic Self*. There are social and cultural factors that influence genes and synapses, and these in turn affect our behavior and our sense of who we are.

We rarely distinguish between our core self—the "I" of basic consciousness—and personal identity, a complex, multilayered entity that shifts and evolves throughout our lives. And we all have multiple facets to our personal identity: the inner private self and the social self, the conscious self and the unconscious self, the physical self, and, in the Internet age, the virtual self. Somehow all those aspects come together to form an integrated, conscious, and cogitating human being unlike anyone else.

How could a humble science writer like me make any sense of something so elusively complex, when the world's most brilliant thinkers are still grappling with the precise mechanisms behind how this marvelous integration occurs? "You can't. Why should you?" New York University neuroscientist David Poeppel bluntly asked over lunch in a Greenwich Village deli. "We work for years and years on seemingly simple problems, so why should a very complicated problem yield an intuition? It's not going to happen that way. You're not going to find the answer." But, he added kindly, perhaps responding to my crestfallen face, that doesn't mean one shouldn't try to wrestle with the fundamental questions. "There are problems and there are mysteries, and problems can be solved," Poeppel said, even if those solutions are far in the future.

I had somewhat naively assumed this book would be a lighthearted romp through genotyping, a brain scan, and a few personality tests, after which I could pinpoint with triumphant accuracy exactly which traits could be attributed to my genes and which were due to my environment, or some combination thereof. Instead, I found myself scrambling to navigate bumpy empirical ground that was constantly shifting beneath my feet, veering dangerously close at times to the precipice of philosophy. To make the task a bit more manageable, I chose to break down the self into component parts to shed light on how the pieces work, individually and collectively. Part I explores "me" (genes, synapses, personality traits), while Part II focuses on "myself" (behaviors, identity), before Part III plunges into the "why"—the murky meta-depths of how human consciousness emerges from all those fundamental processes, and how we, as conscious creatures, craft our personal narratives.

This is the most personal book I've yet written. A book exploring the science of self could hardly be otherwise. Yet such an approach is not without risk of the dreaded overshare. The sixteenth-century essayist Michel de Montaigne once cautioned against too much self-revelation: "Don't discuss yourself, for you are bound to lose; if you belittle yourself, you are believed; if you praise yourself, you are disbelieved." Wise words. But he didn't heed his own advice: instead, he pioneered an entire genre, the personal essay, and spent years chronicling the minutiae of his life. He was an inveterate blogger centuries before there were blogs, scribbling enthusiastically on topics as diverse as smells, clothing, drunkenness, idleness, politics, imagination, and even his cat with colloquial flair, thereby demonstrating the value in using the quirks and foibles of an individual to illuminate more general truths.*

"I study myself more than any other subject," Montaigne

---

* "When I play with my cat, how do I know that she is not playing with me rather than I with her?" he asks in "An Apology for Raymond Sebond."

boasted. "That is my metaphysics, that is my physics." He had a point. We are all metaphysicians at heart, natural-born story-tellers, and our favorite protagonists are ourselves. Whether the objective is personal, intellectual, or some combination thereof, the narrative always involves a journey. In the end, there may be more questions than answers, and we might not always like what we find. That's not the point. Ultimately, the story is not about the destination. It's about everything learned along the way.

I

ME

1

What's Bred in the Bone

It is the common wonder of all men, how among so
many millions of faces, there should be none alike.
—THOMAS BROWNE, *Religio Medici*

S*nnnerk-haaaAACKpt!* Was that the sound of a distressed
cat trying to expel a particularly stubborn hairball? Alas,
no, it was just me, preparing to spit into a small plastic tube
provided by the commercial genetic testing company 23andMe
as the first step in genotyping my DNA. It was not a pretty sight.
Or sound.

Human beings produce between 0.75 and 1.5 liters of saliva
every day, yet it was surprisingly difficult to fill the 2.5 ml tube.
I spent a good five minutes horking and spitting before the level
of sputum finally reached the magic line etched on the tube.
Then I popped the little blue cap on the top of the tube and
shook it vigorously, per the instructions, mixing the saliva sample with a special solution designed to stabilize my DNA for the
bumpy ride through the U.S. postal system. Eventually it
wended its way to a lab, where it was analyzed by a bevy of lab
technicians in crisp white lab coats and plastic goggles as edgy

alternative music played in the background—because aren't all sequencing labs staffed by colorfully eccentric characters from *CSI*? If all went well, soon I would have at my fingertips the secret code to unravel the mystery of why my eyes are green and not brown, why I loathe Brussels sprouts, and whether I have wet or dry earwax.

Personal genotyping is all the rage these days. Human beings are curious by nature, never more so when the subject is ourselves, and we now have a tool at our disposal capable of giving us unprecedented access to our genetic data. How could we not be fascinated? Even the 5,300-year-old Ötzi the Iceman—a mummified body found frozen in the Italian Alps by a couple of hikers in 1991—has had his genome sequenced, whereby the world learned that he likely had brown eyes, type-O blood, blocked arteries, Lyme disease, and lactose intolerance. (If mummies could blush, he'd be mortified, I'm sure, to discover his personal details plastered all over the Internet.)

My interest went beyond mere curiosity about my ancestry, although the results did confirm what I already knew about my ethnic heritage: predominantly Norwegian, with a motley mix of other Northern European strains. As an adopted child, I have zero information about my immediate biological family history beyond the broadest generalizations. That includes medical history, a critical component when doctors evaluate patients for diagnostic purposes. After a lifetime of answering medical history questions during exams with a dismissive shrug—"I'm adopted, so I really don't know"—23andMe's analysis could help shed some light on my genetic risk factors, although it would not substitute for a detailed family history.

Providing access to such information is one of the guiding principles behind the company's mission statement. It's not just

about using the tools of genetics research to gain a deeper understanding of how genes shape our lives. It's also about enabling
individuals to control their own genetic information, and to be
fully engaged participants in the ongoing ethical, social, and policy debates that such commercial technologies inevitably spark.

A few weeks after receiving my test results, I found myself at
23andMe's headquarters in Mountain View, California, sitting
across from cofounder and CEO Anne Wojcicki. The daughter
of Stanford particle physicist Stanley Wojcicki, she received her
undergraduate degree in biology and spent a decade as a highly
successful hedge-fund manager, specializing in health-care and
biotechnology companies, before deciding to join the front lines
of a budding revolution.

In 2005, Wojcicki attended a swanky dinner party in Mon-
terey. With characteristic frankness, she unabashedly quizzed
her tablemates about their urine. She wanted to know which of
the well-heeled guests could smell asparagus in their urine when
they ate the vegetable—a peculiar sulfurlike scent that results
when asparagus is digested. At least, that's the case for most
people; a few folks at the table had no idea this happened, because they had a genetic variation that kept them from detecting
the smell—either that, or they didn't make a habit of sniffing
their own urine, or wouldn't publicly admit to doing so.

That was Wojcicki the curious biology major talking, but
the hedge-fund manager spotted an untapped opportunity for
bringing the power of genetics out of the laboratory and into the
hands of consumers. She knew that pharmaceutical companies
were already exploring personalized medicine, based on the explosion of genetic information being compiled by scientists. But
average people don't have access to the same data, and probably
wouldn't be able to make much sense of it even if they did.

"Everybody talks about the patient, but the patient doesn't really have a voice," Wojcicki said of her rationale. "If we really want to make significant changes in our own health care and our own destinies, we need everybody coming together." In other words, it's not just about the "me," however self-indulgent it might feel to order a personal genotype. With cofounder Linda Avey, she developed the concept of a single database that would compile all the available research and update it as new discoveries were made. Serendipitously, Wojcicki happened to be dating Google CEO Sergey Brin (now her husband), who also saw the commercial potential, and stepped in as an "angel investor" to help get the fledgling venture off the ground.

The timing couldn't have been better. We are still in the early days of the personal genomics revolution, made possible by the completion of the $3 billion Human Genome Project in 2003. Every day brings news of another organism's genome being sequenced—a phenomenon science writer Carl Zimmer has cheekily dubbed Yet Another Genome Syndrome (YAGS)—or a new study identifying a genetic factor underlying some trait. Even the *New England Journal of Medicine* has compared keeping up with all this new research to drinking from the fire hose.

At the same time, the technology for genome sequencing has been advancing at an exponential rate, with an inversely falling price point. It cost $1 million to sequence the genome of James Watson, who first discovered the structure of DNA with his colleague Francis Crick. By 2007, when geneticist Craig Venter had his genome sequenced for a second time, the cost had dropped to $300,000. In 2012, Oxford Nanopore Technologies announced a disposable DNA sequencer the size of a USB memory stick for $900 a pop, as well as a larger bench-top version. Put twenty of the latter devices together and it is possible to sequence an entire human genome in fifteen minutes—or so the company claimed. That same year, Life Technologies unveiled a prototype bench-top sequencer it claimed could decode a human genome in one day for less than $1,000. Whether or not those claims are borne out, the day when genome sequencing becomes so affordable as to be commonplace is closer than you think.

A genotype consists of all the genes that influence traits, while phenotype describes how those traits are expressed statistically through multiple generations. Your unique DNA sequence is your genotype; all your traits taken together make up your phenotype. 23andMe's analysis does not map out your entire genome. Rather, the process identifies known genetic variants you possess and gives a statistical analysis for certain key traits or disease risk factors based on the scientific literature. In 2011, 23andMe added an option that sequences just the coding portions of expressed genes to select customers for $1,000. It is still limited in scope, accounting for around 1 percent of the full human genome, but those areas may account for as many as 85 percent of disease-related mutations.

What sets 23andMe apart from its competitors is social networking. When you order the kit, you also sign up for a one-year subscription to what amounts to Facebook for the genotype set. Subscribers can access their results, but they can also choose to share their genome with other users—possibly finding distant relatives in the process. They can also fill out surveys about their lifestyle habits and medical history designed to elicit supplementary information for ongoing scientific studies based on the compiled genetic data, so subscribers actively contribute to ongoing research. As of December 2011, 23andMe boasted a database of more than 125,000 individual genotypes, and that number is still growing by the day.

The company has its fair share of naysayers, too, who cite privacy concerns and the possible misuse of genetic data by insurance companies or other entities—not to mention the fact that the average person may not have sufficient scientific background to understand what their customized report is really telling them.

Wojcicki acknowledged those concerns, and the company has policies in place to address them. Concerned about privacy, or how your data is being used? "Your DNA is yours," she insisted. "You can download your DNA and quit at any time." Confused by all those messy statistics? There are careful explanations about what the numbers mean, other nongenetic risk factors, background information about various diseases and traits, and lifestyle tips to reduce your overall risk, all backed up by academic citations, and wrapped in a user-friendly package with videos, animations, and colorful illustrations. I can personally attest that it's possible to lose yourself for hours on their Web site. And a public better educated about genetics is better equipped to evaluate any proposed new policies and defend its rights accordingly. But to reach that deeper level of understanding, many of us will need to discard the simplistic notions about how heredity works that we learned back in grade school. I'm talking about those pea plants.

Peas in a Pod

In the 1860s, an Austrian monk in what is now the Czech Republic spent several years conducting methodical experiments crossbreeding pea plants. Johann Mendel was the son of peasants, who worked on the family farm and enjoyed gardening and keeping bees. He spent three years at the University of Olomouc, studying hereditary traits of plants and animals. Eventually, Mendel chose to enter a monastery because that was the only option available to a man of his limited financial resources. He changed his name to Gregor when he took orders, and he had the good fortune to land in a monastery with an abbot who loved plants as much as he did. The Abbey of St. Thomas in Brno boasted a large experimental garden, ideal for fostering Mendel's scientific interests.

Brother Gregor was curious about how certain unique traits in plants were passed on to successive generations. He picked pea plants because they could easily be grown in large quantities, and have male and female reproductive parts, allowing them to both self-pollinate and cross-pollinate with other plants. It was a simple matter to control the breeding: Mendel just removed the stamens of any plant he wished to cross-pollinate before it reached maturity.

After carefully cultivating non-hybrid ("pure" or "true") pea plants, Mendel focused on seven easily recognizable traits usually expressed in one of two forms, including flower color (purple or white), pea color (yellow or green), and whether the pea shapes were smooth or wrinkled. He bred many generations of pea plants, over seven years, and carefully recorded the outcomes, compiling notes on more than 10,000 individual plants. Over time, he noticed that traits seemed to be inherited in specific ratios via hereditary units he called "factors" (genes).

In the Mendelian system, one gene, or "factor," may control for one trait, but there may be two different expressions of that trait. For instance, a single gene codes for the color of the peas, but there are two versions—one that codes for yellow peas and one that codes for green peas.* Today we call these alleles. Each parent contributes one allele for a specific trait, so the offspring inherits two copies of the same gene, which may be the same, or code for different variants. In the latter case, the result is a hybrid. Variations in subsequent generations occur because of a random mixing-and-matching process that sorts out which genes from which parents are ultimately expressed in a given offspring.

One of those two alleles in a hybrid will be dominant and one will be recessive. Only the dominant allele will be expressed—that is, affect the appearance of the offspring—but the recessive gene can still be passed on to subsequent generations. Contrary to what was popularly believed at the time, inherited traits don't "blend" when they are passed down to subsequent generations. A pea flower will be either purple or white. It will not be some intermediate color, like pink or lavender.† And each trait is passed down independently of all the others. So just because an offspring plant inherits purple flowers, this does not affect the likelihood of whether it will produce yellow or green peas. That is called the law of independent assortment.

In the 1934 film *The Thin Man*, Dorothy Wynant is about to marry a fine, upstanding young man when her eccentric inven-

---

* "Coding" in this context means the gene's sequence of nucleotides that triggers production of a protein that affects, say, the color of pea flowers. This is distinct from "decoding" the genome, in which one is determining the chemical structure of a DNA molecule and expressing it in terms of letters.

† There *are* some plants with what's known as "incompletely dominant" genes that produce a "mixed" expression. For example, a gene that codes for either white or red flowers that is incompletely dominant may produce pink flowers.

tor father, Clyde Wynant, mysteriously disappears. Then his longtime mistress turns up murdered, and suspicion falls on the missing Wynant. Dorothy refuses to believe it, until confronted with some damning circumstantial evidence of her father's presumed guilt. Distraught, she breaks off her engagement, on the grounds that the family bloodline is tainted and she doesn't want to give birth to a brood of murderers.

Her bookish, socially maladapted brother, Gilbert, overhears the conversation, and attempts to comfort her with science. "You know, you're wrong about all of your children being murderers," he pronounces. "I've studied the Mendelian laws of inheritance and their experiments with sweet peas, and according to their findings—and they've been pretty conclusive—only one out of four of your children will be a murderer. So the thing for you to do would be to have just three children." To Gilbert's credit, he quickly spots the flaw in this plan: "No, no. That might not work. The first one might be the bad one."

Was Gilbert correct that only one in four of Dorothy's children would be murderers? Eccentricity and weakness of character do seem to run in the Wynant dynasty. Mimi, Dorothy's mother, is a vain, conniving creature constantly leeching money out of her ex-husband to support her oily gigolo of a second husband. The oddly detached, emotionally stunted Gilbert admits to a latent Oedipal complex and has a morbid fascination with all the corpses piling up. (Given his ghoulish proclivities, I suspect he grew up to be either a brilliant forensic pathologist or a serial killer.) Even her father, while likeable, has a touch of madness about him. Dorothy is the most well adjusted one in the bunch. It's astounding her fiancé didn't flee in horror the first time he met the family.

Let's first take a closer look at what Mendel discovered.

Each of Mendel's pea plants carried two versions of the gene that controls for seed shape—whether a seed will be smooth or wrinkled. Both parents can have the smooth allele (A), both parents can have the wrinkled allele (a), or one will have the smooth and the other the wrinkled allele. If both parents have the same allele, the outcome is easy: all the offspring will inherit that allele. But if the parents have different alleles, one will be dominant (A) and the other will be recessive (a). Both versions will be passed down to all the offspring (Aa), but only the dominant allele (A) will be expressed. In this case, smooth peas are the dominant trait.

The second-generation pea plants with the (Aa) combination are hybrids, not "pure" varieties. Even if only the dominant allele is expressed, they are still carriers of that recessive allele (a), and hence can pass it on to their offspring. So the odds play out a bit differently for the third generation of pea plants. There are four possible combinations: (AA), (Aa), (aA), and (aa). One out of four third-generation offspring will be (AA), and two out of four will be (Aa), but since (a) is recessive, all three such plants will produce smooth peas. Only the (aa) combination will result in wrinkly peas—a 3:1 ratio or one in four chance.

Now, let's assume that Dorothy's fears are well founded, and Wynant Senior really did kill his mistress. Furthermore, this homicidal behavior is purely deterministic, due to a "murder gene" running in her family; inherit the "murderer" version of that gene, and you will be a murderer. Let's also assume her father carries two copies of the recessive "murderer" allele (mm), while her mother carries two copies of the dominant "non-murderer" allele (MM). Each child inherits one of each (Mm), making them hybrids. Since (m) is recessive, Dorothy and her brother will not be murderers, despite Gilbert's disquieting enthusiasm for dead bodies.*

Dorothy isn't out of the woods entirely. She still carries that recessive gene, and can pass it down to her offspring. It all depends on the genetic profile of the father of those children. If we assume her fiancé is also (Mm), there are four possible genetic combinations for their children: (MM), (Mm), (mM), and (mm). Only one of those (mm) will express the "murderer" variant, so under this scenario, Gilbert is correct. There is a one in four chance that she will give birth to a murderer—unless, of course, her fiancé is (MM), in which case none of their offspring will express the "murderer" allele. In fact, there's only a fifty-fifty chance they will inherit the recessive version (Mm) at all. Conversely, if the fiancé is (mm), not only is he a murderer, but there is now a fifty-fifty chance their children will be murderers too, and a 100 percent chance that their children will be carriers of that recessive trait.

Does that scenario strike you as absurd? It should, because human beings are far more complicated than your average pea plant, particularly when it comes to behavior, and no two people are exactly alike—not even identical twins. Yet that complexity arises from a surprisingly small number of genes: around 23,000, give or take a few thousand. For comparison, yeast has 6,000 genes, a fruit fly has 12,000 genes, and the simplest worm has 18,000 genes—only 5,000 or so less than humans. Furthermore, we share 99 percent of our genes with the rest of the human race. So what sets us apart from yeast or worms, and what

---

*For simplicity's sake, we'll assume both parents are crossbreeding nonhybrids.

accounts for the astonishing diversity among the human population? To answer that, we need to take a look inside this mysterious thing called DNA.

The Double Helix

Deep in a converted castle vault in Tubingen, Germany, a young Swiss physician, just shy of thirty, faced a wicker basket teeming with used bandages covered in pus, courtesy of a local surgical clinic. But Johann Friedrich Miescher didn't see the filth. To him, the bandages were a rich source of white blood cells, ideal for his experiments—if he could just figure out how to wash off the cells without damaging them. The year was 1869, and the centrifuge had not yet been invented. The newly minted doctor had opted to pursue research at the University of Tubingen instead of clinical work, after a nasty bout of typhus left him partially deaf, hampering his ability to treat patients. It would prove to be a serendipitous decision.

When most of us think about DNA, Miescher's name is not the first to spring to mind. We usually associate DNA with James Watson and Francis Crick, who—in conjunction with Rosalind Franklin—first discovered the unusual double-helix structure of the molecule in 1953. But it was Miescher who set the stage for the revolution with an accidental discovery nearly a century earlier. He was interested in the proteins found in white blood cells, and came up with a sodium sulfate solution to filter those cells out of the pus-filled bandages, letting them settle to the bottom of a beaker. When he extracted the proteins from the cell nuclei, he discovered that it wasn't a protein at all, but a new substance we now know as nucleic acid.

Miescher spent years studying the chemical properties of

nucleic acids, but was stumped as to what their function might be, although he suspected they might play a role in heredity. He died in 1895 of tuberculosis. It wasn't until 1943 that an American scientist named Oswald Avery proved that the organic molecule now known as deoxyribonucleic acid (DNA) carried genetic information, and by then Miescher had been largely forgotten.

The molecule he discovered is very much a household name. Yet most of us have a rather fuzzy understanding of how DNA really works. Take a look at the human genome under a metaphorical microscope, gradually increasing the resolution to observe smaller and smaller scales, and a rough hierarchical structure emerges. The genome is made up of DNA molecules, that famous double helix of intertwining strands coiled up inside a cell's nucleus. These in turn are sorted into twenty-three pairs of rod-shaped chromosomes, which contain the genetic information for many different traits. Within each chromosome are shorter segments of DNA, which we call genes, combined into one long strand.

I like to think of the genome as a cookbook and the genes as the recipes containing detailed instructions for regulating how cells function and which traits should be expressed. Those genes are written in a special chemical alphabet consisting of just four letters: adenine (A), guanine (G), cytosine (C), and thymine (T). These letters form complementary pairs—(A) always pairs with (T) and (G) always pairs with (C)—connecting the long strands of DNA. If the DNA molecule is a ladder, the rungs are the three billion interlocking pairs of chemical bases. All the variation among human beings arises out of 23,000 genes, twenty-three pairs of chromosomes, and four chemical base pairs. If this seems far-fetched, think of all the amazing recipes one can concoct out of a small set of raw ingredients. Nature gave us a well-stocked kitchen pantry.

How does DNA get the body to carry out its instructions? The DNA transfers the coded genetic information to another type of nucleic acid, ribonucleic acid (RNA), through a process called transcription. There are many different kinds of RNA, each with its own function within the cell. The RNA in the cell's nucleus sends out another type called messenger RNA (mRNA) to a collection of other RNAs and proteins that together make up the ribosome. This is where the sequence is decoded, and the body then uses those instructions to manufacture proteins. As Crick famously summed it up: "DNA makes RNA, RNA makes proteins, and proteins make us." If we extend the cooking metaphor, think of the genome as the master chef, who carries all the recipes in his head, and collects them into a coded cookbook (the DNA), which is passed onto the sous-chef (RNA). The sous-chef then transfers those instructions to the main kitchen staff (ribosome) via the mRNA, and the cooks at various stations perform their respective functions to create the meal.

There is a flow of information from DNA to protein, but only the DNA is passed down to one's offspring. Only one chromosome from each of the twenty-three pairs goes into a sperm or egg, and these are selected quite randomly—shades of Mendel's random mixing. So every sperm and egg cell is unique. When sperm and egg are joined in fertilization, their single sets of chromosomes combine so that the resulting single cell, and the human who grows from it, has a complete set of twenty-three pairs. Those pairs determine which traits a particular child inherits from each parent. Each father and mother has two full sets of chromosomes in their cells. Everyone shares the same genes, but there are tiny variations within those genes—

the result of mutations—called single nucleotide polymorphisms (SNPs) that make them uniquely ours.

SNPs are a bit like the typographical errors that occur over repeated copying, except in this case it is not a manuscript being copied, but DNA. Cells replicate by dividing in two, and each new cell carries the same full set of "recipes" contained within the DNA. But with each replication, small variations creep into the sequence at particular locations on the genome—much like different cooks will tweak standard recipes to lend their personal touch to a dish—such that (A) becomes (G), or (T) becomes (C). There are around ten million of these SNPs in our genetic code. Often these mutations are so insignificant that there is very little perceptible effect on expression of traits, but some variations—either alone or in concert with other genes—end up influencing your appearance, your response to certain drugs or foods, or your susceptibility to a particular disease.

It is those SNPs that 23andMe targets for its genotype analysis. What happened to my saliva sample once I popped it into the mailbox? Upon arrival at the lab, the DNA was extracted from the saliva and then copied repeatedly to ensure there was enough for a full genotyping. Then the DNA was cut into smaller bits and placed on a DNA "chip" the size of a matchbox. It looks like a standard glass slide, except that millions of microscopic beads have been attached to its surface. Each bead contains a piece of complementary DNA that matches to a corresponding site on the genome where a specific SNP is located, just like a lock and a key. My extracted DNA stuck to those probes that matched whichever version of the specific SNPs I have. When there was a match it glowed, thanks to the addition of a fluorescent marker.

I never gave a second thought to the consistency of my

earwax until now. The gene in question codes for a protein designed to transport certain compounds (like drugs) out of cells. Earwax serves a similar purpose, trapping dirt and dust before it can penetrate the inner ear. So the protein plays a role in secreting the oils that make earwax wet; a shortage of such oils results in dry, flaky earwax.

There are two alleles: those who inherit one version have wet earwax, like me; those who inherit the other version have dry earwax. That gene is close to a simple Mendelian trait, but it has more than one function. It may also play a role in how much colostrum women secrete after giving birth. That's the nutrient-rich substance rife with antibodies that protects newborns until their immune systems are fully developed. So there is no gene specifically for wet or dry earwax. Rather, there is a gene that codes for a protein, and it comes in two varieties, one of which results in wet earwax, while the other results in dry.

It is entirely possible for a trait to be heritable, meaning an individual can pass it down to his or her offspring, and yet to have what amounts to zero heritability—a statistical measure that applies to the population as a whole. Genetic variation accounts for between 80 and 90 percent of the observed variation in height, with the remaining 10 to 20 percent attributable to differences in environment, most notably nutrition. This is a trait that is easily defined and highly heritable, yet there is no single gene helpfully labeled "height." As recently as 2010, scientists had identified fifty small genetic variants related to height, yet these accounted for only 5 percent of the observed variation in the population. No single genetic variant has a particularly large effect. Rather, there is a cumulative effect: innumerable tiny variants acting in aggregate, along with environmental factors, ultimately determine one's height. Matters are even squishier when it comes personality traits, which are roughly 50 percent heritable, or behaviors.

The Personal Genome Project at Harvard University has found, on average, more than 100,000 genetic variants unique to each person who participated in the project; one-eighth of that variation had not been observed previously. Forget everything you've heard about the "warrior" gene, the "gay" gene, the "drunk" gene, or the "optimism" gene. There just aren't enough genes in the human genome for the "one gene, one trait" model to work. "Most traits are affected by many genes and most genes are involved in the development of multiple traits," biologist-blogger Bora Zivkovic has explained. "It is not which genes you have, but how those genes interact with each other during development that makes you different from another individual of the same species, or from a salmon or a cabbage."

Taster's Choice

In the X-Men franchise, a small subset of otherwise ordinary people spontaneously develop special abilities: flying, telepathy, teleportation, rapid healing, manipulating magnetic materials, walking through walls, shooting laser beams from the eyes, and so forth. They are tragic superheroes (or super villains), hated and feared by the rest of humanity. The explanation offered for their condition is a sudden mutation in their genetic code; they are the next leap in human evolution. As a young Charles Xavier—himself a mutant with powerful telepathic abilities—explains in *X-Men: First Class*, "Mutation took us from single-celled organisms to being the dominant form of reproductive life on this planet. Infinite forms of variation with each generation, all through mutation."

What is the mechanism by which this mutation occurs? The franchise is vague about the scientific details, but there is a means by which single-celled organisms might have abruptly developed "super traits" that led to them rapidly evolving into higher life forms—a concept dubbed "hopeful monsters." Geneticist Richard Goldschmidt coined the term in *The Material Basis of Evolution* to explain how nature managed to bridge the gaping chasm between microevolution and macroevolution.

Goldschmidt didn't think the accumulation of small, gradual changes over time—the more accepted mechanism for genetic mutation—was sufficient to account for big evolutionary leaps forward. He argued that the results of most mutations would be "monstrous," and would die before producing offspring and passing on those particular traits. As an alternative, he suggested that once in a while, a single mutation would produce a very large effect—rather than an aggregate of tiny mutations over time—resulting in a trait highly beneficial to surviving in a specific environment, giving the creature with that trait an evolutionary advantage. He called this hypothetical creature a "hopeful monster" that would then pass on the trait to its offspring, founding a new lineage.

Biologists didn't much care for this notion when Goldschmidt introduced it, although the discovery of regulatory genes in the 1970s lent credence to his ideas. It required a significantly different concept of how evolution occurs—not the nice, smooth continuum of small changes over long periods of time, but more of a herky-jerky process, in which continuity is occasionally disrupted by large, rapid mutations. There is evidence in the animal kingdom of the "hopeful monster" phenomenon: the loss of feathers in a certain species of vulture, for example, or leg bristles on some species of fruit fly. So changes to a single gene can have big effects on expressed traits. It's just that scientists can't be sure if it was due to a single mutation, or a series of smaller mutations adding up over time.

The notion is appealing enough that I wondered if I could be a hopeful monster, one day taking an evolutionary leap forward to develop my own special powers. I was hoping for teleportation. Alas, my 23andMe results showed no evidence of a mutated "X" gene. Instead, I have the far less impressive ability to taste bitterness—an ability I share with roughly two-thirds of the population, which frankly makes me feel even less special. There is not much call for a superhero dedicated to protecting finicky children from broccoli.

In 1931, a chemist named Arthur L. Fox accidentally released the powdered form of phenylthiocarbamide in his lab. He didn't notice anything unusual, but his lab mate sensed a bitter taste. Subsequent experiments confirmed that this variation existed in the broader population, and that not being able to taste bitterness was a recessive genetic trait. Those who can sense bitterness are probably responding to compounds called glucosinolates, present in most cruciferous vegetables, like broccoli, Brussels sprouts, and cauliflower. I happen to dislike all three. At least part of that lies in my genes.

About 25 percent of the population can't taste propylthiouracil, a chemical that is similar to the bitter compounds found in cabbage, raw broccoli, coffee, tonic water, and dark beers. They are, in essence, "taste blind." I do not fall into that 25 percent, thanks to a gene that encodes taste receptors on the tongue. I have the (GG) variant; that and the (CG) version both result in being able to taste bitterness, since (G) is the dominant allele. The (CC) allele is the taste-blind version, although even then, there is a 20 percent chance that you still might be able to sense some bitterness, depending on other genes you inherit. There are twenty-five "bitterness" genes known thus far. Different bitter foods act through different receptors, and people can be high or low responders for one but not another. This is probably why I don't mind grapefruit or tonic water, but balk at cruciferous vegetables.

While broccoli and cauliflower taste mildly unpleasant, fresh raw tomatoes make me nauseated. Once, when I was a child, my mother lost patience watching me shove tomatoes to the side of my plate, and insisted I couldn't leave the dinner table until I ate them. I put it off as long as possible, but finally, desperate to get away, I shoved the offending food into my mouth—and promptly gagged and spit it up. My mother, to her credit, threw up her hands in resignation. Her daughter would never eat raw tomatoes. This visceral dislike was incomprehensible to her; she loves them as much as I despise them.

Frankly, it's a little incomprehensible to me: I like ketchup, salsa, and marinara sauce, provided there are no huge chunks of tomatoes. Boil and puree those suckers, and season with tons of garlic, olive oil, basil, thyme, and oregano, and it overcomes even my rebellious palate. Traditional home cooking techniques— long, slow simmering times, reheating sauces day after day— do seem to change tomatoes by altering an antioxidant called lycopene. The chemical properties of the molecule are the same whether cooked, processed, or raw, but the form of lycopene that shows up in the human body has a bent shape, while the form in your standard raw tomato is more linear. This limits how much of the nutrient can be absorbed into the bloodstream. Subjecting tomatoes to intense heat, and combining them with fat (like olive oil), changes the shape of the lycopene molecules from linear to bent so that it is more likely to be absorbed by the body.

But my aversion is more likely due to one of the more than four hundred flavor compounds in raw tomatoes, although scientists have yet to identify any specific genes. People like me—there are more of us than you think—might just lack certain taste receptors, preventing us from appreciating the rich, sweet, meaty flavor of raw tomatoes that the rest of you are always rhapsodizing about.

That is almost certainly the case with cilantro. I love cilantro. To me, it tastes fresh and citrusy with just a tinge of a grassy note. But to many people, it tastes like soap. Celebrity chef Julia Child confessed in a 2002 Larry King interview that if she spotted cilantro in her food, she would pick it out and throw it on the floor. Cilantro haters can blame their genes. Behavioral neuroscientist Charles J. Wysocki asked forty-one pairs of identical twins and twelve pairs of fraternal twins to rate their reactions to the taste of cilantro, on a scale of plus eleven (*yummy!*) to minus eleven (*gross!*), with zero indicating a neutral response. More than 80 percent of the identical twins rated the taste of cilantro on a par with their siblings, compared to just 42 percent of the fraternal twins, suggesting a genetic link.

From a chemical-compound perspective, cilantro is less complex than raw tomatoes. According to food chemistry expert Harold McGee, there are around six substances that contribute to the telltale aroma of cilantro, mostly fat molecules known as aldehydes. Similar aldehydes can also be found in soaps and lotions—and bugs, which make use of aldehyde-drenched body fluids as either an attractant or a repellant. In contrast, Wysocki says that the fresh, flavorful, pleasantly herbal compound derives from dodecenal. He thinks those who hate cilantro are reacting to its odor more than its flavor, and that the haters can't detect the pleasing chemicals in the leaf. Instead, they just detect that soapy aspect.

Wysocki conducted a gas chromatography experiment with a device that uses heat to separate the various molecules in cilantro, so subjects could take a whiff of each separate compound. People who liked cilantro first detected the soapy scent, followed by the stronger citrusy, herbal scent we savor; but cilantro haters couldn't smell the latter. "It's possible that they have a mutated or even an absent receptor gene for the receptor protein that would interact with the very pleasant-smelling compound," Wysocki told MSNBC in 2011.

Several genetic variants may contribute to cilantro taste preferences, and in 2012, 23andMe identified one potential culprit that influences sensitivity to aldehyde. However, the variants identified thus far collectively account for less than 10 percent of cilantro preference in the general population. We still know very little about the genetic underpinnings of our taste preferences, certainly not enough to predict anyone's reaction to cilantro (or raw tomatoes) based solely on their DNA. Even determining eye color is not as straightforward as you might think.

The Eyes Have It

In *X-Men: First Class*, Xavier is at a pub celebrating earning his PhD when he attempts to seduce an attractive young woman with mismatched eyes by dazzling her with his knowledge of genetics. "Heterochromia is in reference to your eyes, which I have to say are stunning," he purrs. "One green, one blue. It's a mutation. It's a very groovy mutation."

With all due respect to Xavier, it's not always a mutation. Occasionally it is the result of trauma to the eye, as in the case of rock star David Bowie: one of his eyes changed color when he was struck during a fight. Another possibility is that a particular pigment gene is turned on in one eye, but not expressed in the other. But it's true that heterochromia can also be the result of a mutation, which might occur in the gene of just a single cell, making it genetically distinct from the rest of the body—a phenomenon known as somatic mosaicism. If that gene happened to be associated with eye color, it could result in one differently colored eye.

Like height, eye color is highly heritable (between 90 percent and 99 percent), but there are not separate genes for blue, green, and brown eyes—or combinations thereof, resulting in the astonishing range of hues found in people all over the world. Eye color is a "multigenic" trait. That means there are multiple genes, with multiple variants, interacting with one another in complicated ways to determine eye color.

Your eye color depends upon how much melanin is produced by the cells in your iris, the same pigment that gives color to hair and skin. Different eye colors arise because there are different amounts of melanin in the outer layer of the iris, as well as differing ratios between two types: eumelanin (a blackish-brown pigment) and pheomelanin (a reddish-yellow pigment). The darker your eyes, the more total melanin your irises produce, because there is more eumelanin, which absorbs light, making the eyes appear brown. Lighter eyes have less melanin and a higher percentage of pheomelanin; they absorb less light. Instead, light passes to the deeper layers of the eye, where it is scattered by proteins, and then reflected back through the iris, giving it a blue color. Green or hazel eyes are lighter variations of brown eyes.

But where does all that pigment come from? It is linked to a gene that encodes a protein that alters the pH of iris cells that produce melanin. That accounts for 74 percent of variation in eye color; other genes play a role, but most have yet to be identified. The 23andMe analysis is based on an SNP in a neighboring gene, which may influence how much of that critical protein is produced by the cells in the iris. People with two copies of the (A) version of this particular SNP will almost always have brown eyes. One copy of the (A) version and one copy of the (G) version can produce either brown or green eyes, depending on how it interacts with other genes. Those who, like me, have two copies of the (G) version will usually have blue eyes, although interactions with other genes create a small probability of green eyes, which won out in my case. My genotype results predicted a 72 percent chance that I would have blue eyes, a 27 percent chance that my eyes would be green, and only a 1 percent chance my eyes would be brown.

Alternatively, the young woman who caught Xavier's interest might be a chimera—a very rare creature indeed. The Chimera was a mythical fire-breathing monster described in Homer's *Iliad* as having a lion's head, goat's body, and serpent's tail. But there is also a rare genetic condition called chimerism, in which one person has two distinct sets of DNA. There have been fewer than forty reported cases.

One of the most famous is Lydia Fairchild, who separated from her husband while pregnant with her third child, and took a DNA test to prove her husband's paternity of her two older children. He was the father, but according to the test, she wasn't the children's mother, and Fairchild was prosecuted for fraud. She was only exonerated when she gave birth to her third child, with a witness present to take blood samples from mother and infant for testing. Those DNA tests revealed she wasn't the mother of that child either. Yet a court-appointed eyewitness watched her give birth. If she was a fraud, she was a damned good one, on a par with the world's best illusionists.

Then there is Karen Keegan, a Boston-area teacher who needed a kidney transplant in 1998. She had three grown sons who were tested to see if they could be donors, but the DNA showed that two of them weren't her biological children. This time doctors did additional testing on Keegan, drawing samples from other areas of the body, and discovered she had two sets of cell lines with two separate sets of chromosomes—a mix of two individuals, fraternal twin sisters who fused in the womb and developed into a single infant. Fairchild's lawyers heard about the case, and arranged for their client to undergo more testing as well. She, too, turned out to be a chimera. The DNA in Fairchild's skin and hair didn't match that of her children, but the DNA from her cervix did.

Chimerism might not be as rare as previously believed; some researchers are beginning to think there might be a little bit of the chimera in all of us. Most cases simply aren't detected. There aren't many outward signs or symptoms, but eyes of slightly different coloration are one of the most common indications. Perhaps the X-Men are like chimeras: seemingly rare genetic freaks of nature that cause "normal" humans to recoil in fear and horror, until we discover there's a little bit of the mutant lurking in our own bodies, a side effect of our genetic diversity. It is all those tiny changes at the genetic level that help make each one of us unique. Those are the groovy mutations. Yet as the X-Men would be the first to caution, there are risks as well as benefits to mutation.

Doctor in the House

The TV series *House, M.D.* showcased a veritable rogue's gallery of the most obscure, rare afflictions each week, bound to send any self-respecting hypochondriac scurrying to the Internet after every episode to look up the disease du jour. One of the most powerful story arcs concerned Thirteen (née Remy Hadley), a brilliant young woman on House's medical team who discovered she had inherited the gene for Huntington's disease, a neurological disorder that had claimed her mother years earlier. It is not a pleasant way to die: the deterioration drags on for a decade or more as the patient loses motor control, memory, and brain function, eventually ending up bedridden, unable to speak, move, or even swallow, until death finally brings release. It afflicts between 25,000 and 30,000 people in the United States alone.

Huntington's is caused by a defect in a particular chromosome, one of the first genes to be identified in 1983. Anyone with the defect eventually will develop Huntington's, and there is a 50 percent chance that each child will also inherit the disease. An anonymous British war correspondent watched his father and brother succumb to Huntington's and described his trepidation awaiting his own test results in the *Guardian* in 2009: "It was a black-and-white test. Either I would be fine, and my children would be fine, or I would die the most horrible death in lingering misery." Alas, he found that he would share in their fate, although his daughter would be spared: she did not inherit the defect from him.

The prognosis really is that black and white: if you have that genetic defect, you will succumb to Huntington's, unless something else gets you first. It's a question of "when," not "if." But

95 percent of known diseases or conditions don't have such a neat one-to-one correlation. Instead, there are genetic risk factors. Whether you develop a given disease is only partially dependent on genetic predisposition, and often involves many different genes interacting with one another and countless other variables. Genes might be deterministic, but they are not destiny. The human body is a complex system. Even with an elevated risk for a specific condition, factors like family history, lifestyle changes, and other as-yet-undiscovered genes might tip the odds in one's favor.

My 23andMe results came with a handy primer on how the company calculates those odds. First they consider the average incidence: how often a given condition occurs in the population at large, over a carefully defined period of time. So if twenty-five out of one hundred people will be diagnosed with a condition, there is a 25 percent chance of any member in the general population developing that condition. Next, the analysis looks at just those people with a specific genotype and determines what percentage of those are likely to be diagnosed with the condition over the same specified time period. That percentage is then compared to the average incidence rates in the general populace.

Naturally, there are nuances. Age is often a factor. I was surprised to find I had an elevated risk of Type-1 diabetes. My odds are 14.6 out of 100, compared to 1 out of 100 in the general population. But this condition tends to develop at a young age; the risk drops significantly as one matures without becoming diabetic. When I readjusted the sliding bar to reflect my current age, my odds dropped to 1.7 out of 100, compared to 0.11 out of 100 for the population at large.

Some results on the 23andMe site are reported with high confidence—meaning they are based on well-established studies—

while others are still preliminary, with only one or two small studies conducted so far. The two markers included for Parkinson's
disease are based on preliminary research, specifically a Web-
based survey comparing 3,426 people with the disease to more
than 29,000 people without it, all of European descent. People
with the (CC) marker were 1.2 times more likely to develop Parkinson's—including me. That's the bad news. The good news is
that I inherited the (TT) version of the second marker, giving me
typical odds compared to the general population. This balances
out the influence of the other marker to some extent.

As for Alzheimer's disease, I have a decreased risk, based on
more established research. It's between 60 to 80 percent heritable. Roughly 7.1 out of 100 people in the general population are
at risk for Alzheimer's; my risk is estimated at 4.9 out of 100,
even beyond age seventy-five (most people are diagnosed after
eighty). However, there are at least eight other markers that influence the risk, albeit with weaker effects.

Huntington's, Parkinson's and Alzheimer's disease are such
terrifying prospects for many people that 23andMe seals those
reports and makes you confirm that you wish to view them.
This raises the specter of how much the wide availability of
genetic testing might contribute to increased anxiety, particularly for those lacking sufficient scientific background to appreciate what the results really tell them about their genetic risk
factors. As science writer Ed Yong has cautioned, "Some folks
are going to treat their report as nothing more than a sophisticated horoscope, given extra weight by the specter of genetic
determinism."

Geneticist Misha Angrist, who had his full genome sequenced as part of the Personal Genome Project, is more opti-

mistic. "I think most people understand that they are not buying a crystal ball, but, rather, just a snapshot of what we think we know today, the interpretation of which is subject to change," he opined in the *Los Angeles Times*. "I suspect that the public is not as stupid as the paternalists think they are." Angrist pointed out that people deal firsthand with the uncertainties of health risks and medical information all the time and argued that direct-to-consumer genetic testing, in the long run, can help people develop a more positive attitude toward genetics: "We are less likely to be governed by fear and we are more apt to understand that genes are not destiny."

It might come down to personal preference. *House*'s Thirteen balked initially at even being tested for Huntington's disease, telling House she preferred to remain in blissful ignorance, because once you know something like that, you can't go back. Once diagnosed, she struggled to adjust to her new reality, both psychologically and physically, and even briefly wound up in prison for helping her brother—also afflicted with Huntington's—end his life. House, in contrast, thought it was much better to know, because you can't work the problem unless you face the facts.

I, too, prefer to face the facts, even if the prospect is terrifying—like breast cancer. More than 230,000 women were diagnosed with invasive breast cancer in 2011, according to the American Cancer Society, and nearly 40,000 women died from the disease. Women are coached on proper techniques for breast self-exams, and the importance of regular mammograms after age fifty, because survival rates are so much higher when doctors find the disease early. Many of us have experienced that cold, sick dread in the pit of the stomach when we feel an unmistakable lump one morning, and the nail-biting anxiety as we wait for test results to learn our fate. In my case, the lump turned out to be a harmless cyst, but it drove home how little I know about my biological family medical history, and hence my genetic predisposition.

On that score, the 23andMe analysis is less helpful. The major known genes associated with breast cancer (BRCA1 and BRCA2) aren't included in 23andMe's analysis, because until June 2013, another company, Myriad Genetics, held exclusive patents for its own diagnostic products, so 23andMe tests for a mutation on a different gene.* Women with one copy of this mutation are twice as likely to develop breast cancer, compared to those with none. Based on this limited analysis, I have slightly decreased odds of breast cancer: 10.4 out of 100, compared to 13.5 out of 100 for the general female population. By age fifty-five that risk factor drops to 1.2 out of 100, rising slightly again by age seventy-five to 1.9 out of 100.

That doesn't mean I will never develop breast cancer; plenty of other genetic and environmental factors also come into play. And this particular variant is specific to those of Ashkenazi Jewish descent. Perhaps the most common criticism of the service is that it is skewed toward white European races, and sometimes the genetic variants apply to only very specific populations, as in this case, so it is important to read the fine print. If you are African American or Indian or Asian, 23andMe's genotyping will be of less value to you, although this is due partly to the fact

---

*In June 2013, the U.S. Supreme Court ruled that Myriad Genetics could not patent the BRCA1 and BRCA2 genes, effectively ending the monopoly. Mary-Claire King, the geneticist who discovered the BRCA1 gene, told *New Scientist* she was "delighted" by the decision.

that most large studies have been done with white European participants.

One of my best friends, Shari Duffin, lost her mother to breast cancer. At the time, the cost for BRCA testing was prohibitive—around $6,000—and not covered by her insurance, so she opted not to be tested. But a few years later, her annual mammogram revealed two small lumps near the breast-bone. Her mother had been diagnosed with two separate types of breast cancer in her forties, so Duffin wasn't keen on waiting six months for a follow-up scan. "Six months for my mom meant the difference between a stage-one and a stage-three cancer," she said. "There was no way I was going to risk that."

Her insurance still didn't cover BRCA testing, but by then the cost had dropped by half. Duffin, like me, prefers to have as much information up front as possible when making health-care decisions, so she opted to pay for the test out of pocket, although she convinced her insurance company to cover a breast MRI. "I never looked at the [genetic] test as an oracle, only one factor to be considered, but it's an important factor," she said. The news was good: she tested negative for the two most common genes. The lumps turned out to be cysts, and she now has a very accurate baseline image of her breast tissue for comparison on future screenings. That doesn't mean she will never develop breast cancer—just that her genetic risk is reduced.

Stanford physicist JoAnne Hewett was less fortunate: her MRI revealed stage-2 breast cancer, and she faced multiple biopsies, surgeries, and aggressive chemotherapy, followed by five years of endocrine therapy, suffering "every side effect in the book," including losing her waist-length red hair. Today her hair has grown back and she is cancer-free, but still worries about her risk for ovarian cancer, a common metastasis for breast cancer.

Breast cancer is very prevalent on her mother's side. At last count, two cousins, an aunt, and her grandmother's sisters, along with Hewett herself, all developed the disease. Yet those who were tested did not test positive for the most common BRCA mutations. "Genetic testing is in its infancy, in that we hardly know all the genes that could be responsible," she said. Her concern is that those who test negative could be lulled into a false sense of security, ignoring other risk factors. "I am sure that most folks do not understand the complexities involved," she said. "I think it's important to know one's risk factors so that you keep current on screening. But I would hate to live life with fear hanging over me." Granted, she must still face fears of recurrence, but insisted, "One of the gifts of having had cancer is that you appreciate life in a completely different way."

When Craig Venter was asked, in 2011, what he and Watson gained by having their respective genomes sequenced, he quipped, "You probably wouldn't suspect this based on our appearances, but we are both bald, white scientists." Similarly, you could argue, fairly, that knowing my genotype told me very little about who I am, merely verifying the genetic basis for traits I already knew I possessed: ethnicity, eye color, earwax consistency, the ability to taste bitterness, and so forth. It didn't accomplish much more with regard to my health, apart from alerting me to a few heightened risks. But even that little bit of extra information will influence my health decisions for the better. The genotyping also gave me the chance to explore firsthand just how complicated genetic inheritance can be. I still stand with Dr. Gregory House on this one: it's way cooler to know.

2

Uncharted Territory

If the human brain were so simple that we could understand it, we would be so simple that we couldn't.

—EMERSON M. PUGH

"You're not prone to claustrophobia, are you?" the lab technician asked as I laid down on the long flat platform of the doughnut-shaped MRI machine. I assured him I would be fine; this wasn't the first time I'd been inside such a machine. But the setup could prove alarming to someone who feared small, closed-in spaces, or had watched one too many medical dramas in which an MRI malfunctioned, with horrifying results. He strapped my head into a foam-padded plastic cage, since even the slightest movement could disrupt the scan, and slid the platform forward until my head was just instead the machine. I wore headphones to block out the loud *chunk-chunk-chunk* of the machine in operation, my eyes covered with goggles that were linked to an overhead monitor. My right hand rested on a pad with two response buttons, while my left held a panic button—just in case.

I was not here for medical reasons, but for science: my brain

45

was being scanned as part of a cognitive research project in the Houston laboratory of neuroscientist David Eagleman. With his boyish features, hipster sideburns, and infectious enthusiasm, Eagleman is a bit of a media darling, having made a splash a few years ago when he decided to test the subjectivity of "brain time"—why time seems to run more slowly when you are terrified, for example—by strapping LED chronometers onto the wrists of his subjects and tossing them off the 150-foot platform of a free-fall ride at the Houston Six Flags theme park. He needed a convincingly terrifying simulation that was also safe, and the roller coasters just weren't scary enough. I would have volunteered eagerly for such an experiment, but my own small contribution proved far less pulse-pounding. My job was to lie as still as possible and perform simple cognitive tasks while the MRI machine monitored my brain activity.

Eagleman runs numerous research projects ranging from temporal perception, free will, and synesthesia (one of his graduate students is a synesthete herself), to visual perception, and group social dynamics. He has amassed a diverse, highly interdisciplinary group, even poaching an eager young graduate student from the electrical engineering department to explore transference of sensory perception—people who can "see" with their tongues, for instance. The research assistant running my brain scan was Don Vaughn, a tall, athletic California transplant with undergraduate degrees in physics and economics, and an admitted penchant for picking locks, just for larks.

Since I was technically a subject, nobody told me what was being investigated in this particular study, although a heading on one of the forms said it involved pain and memory. This was likely deliberate misdirection, since you don't wish to bias participants beforehand, thus skewing the results. Cognitive neuroscientists are sneaky that way. The pre-test featured questions relating to religious and political beliefs, as well as empathy. Do I identify with the characters in novels? Yes, especially if it is Scout in *To Kill a Mockingbird,* D'Artagnan in *The Three Musketeers*, or Harriet Vane in Dorothy L. Sayers's Peter Wimsey mysteries. Do I cry easily at sad movies? Please—I sobbed like a baby during *Wall-E* when it seemed the plucky little robot might not reboot. Don't even get me started on *Terms of Endearment.*

Once inside the big white doughnut, I was treated to a succession of flashing video images of a hand, laid flat on a table, either being poked gently with a Q-Tip or being given an injection with a long needle. There were several different hands, each with a different colored wristband. Then the whole sequence was repeated, with a twist: each hand was given a specific religious affiliation. Every now and then, a screen popped up with a query—"What religion was the hand you just saw?"— and I dutifully pushed the button corresponding to the correct answer. It was all rather Dada-esque; I kept expecting to see surrealistic collages of rotting fruit and eerily warped faces interspersed randomly with the religious hands.

While all this was going on, the MRI machine was working its magic. The technique is known as functional magnetic resonance imaging (fMRI). Unlike conventional medical MRI, which creates a static image of the brain similar to an X-ray, fMRI monitors the brain in action. When enough neurons fire together in response to a given stimulus, blood flow increases to those parts of the brain involved in processing that input. The fMRI detects this as slight increases in blood oxygenation levels—the so-called BOLD response—in those different regions. In this case, the stimuli are the bizarre video images, and my button-pushing in response. Computer algorithms crunch all that raw data to create a very detailed, high-resolution, color-coded 3-D map of the brain, comprised of thousands of tiny blocks called voxels—like pixels, only pertaining to volume. This belabored statistical process ideally enables scientists to pinpoint which regions of the brain are engaged during specific tasks.

I didn't get to see the final full-color product; that would have to wait for the conclusion of the study. But I did get a quick tour of a black-and-white static image of my brain. Vaughn showed me how to zoom in and out of specific regions on my scan by clicking the mouse to get a split screen with close-up views from different angles. I checked out my frontal lobe, hippocampus, and amygdala, and then moved the cursor to a black blob right between the eye sockets. "What's this big empty space here?"

"Um . . . that would be your sinus cavity." Vaughn assured me that empty sinus cavities are a good thing. I asked if he could tell when a subject had a cold and he said yes, absolutely. One poor soul came in at the height of allergy season, and the scan revealed a sinus cavity that was packed solid with mucus. It couldn't have been comfortable.

Since the first fMRI images were taken more than twenty years ago, it has become one of the most popular brain-imaging techniques among neuroscientists and cognitive psychologists. It is equally popular with the media, because of those eye-popping full-color images—what many people erroneously believe to be snapshots of the brain in action, showcasing the "God spot," for example, or innumerable variations of "this is your brain on <politics, alcohol, sex>," based on which regions "light up" during the scan.

That wording can be a bit misleading, since the resulting image is technically a colorful visualization of statistical data. To create them, neuroscientists run a series of statistical tests, comparing responses in every brain area to two or more tasks. But who wants to look at table after table of raw numbers? Instead, they correlate those numerical results to a color scale, and then superimpose those colors onto an anatomical image of the brain.

Each scan can have as many as 50,000 data points. Computer algorithms average the results from many different scans of participants performing the same tasks—usually one control task and one directed task designed to measure a specific attribute—and then perform a point-by-point comparison of each person's BOLD response per task. The system must take into account the difference in blood oxygenation levels between the control task and directed task; the larger the difference, the stronger the BOLD response, and those that cross the required statistical threshold indicate a correlation between the directed task and the affected brain regions. In the final processed image, that response shows up as a spot of light in the relevant brain region. Even so, false positives can occur: the same brain area lighting up in two different scans because of confounding factors—namely, differences between the tasks that are unrelated to the question the researchers are attempting to answer. In such a case, correlation is not a good indicator of causation.

Perhaps the most colorful example of a false positive was a 2009 paper reporting brain activity in a dead salmon, conducted by neuroscientist Craig Bennett while he was a graduate student at Dartmouth College. It started out as a game to see what weird everyday objects they could scan as part of a routine calibration of the machine, which is usually done by scanning a balloon filled with mineral oil. First, he and his lab partner tried scanning a pumpkin, followed by a Cornish game hen. Then Bennett suggested they scan a whole fish to get better contrast, and duly picked one up early one Saturday morning at the local supermarket. "The clerk behind the counter was a little shocked to be selling a full-length Atlantic salmon at six thirty a.m., especially when I told her what was about to happen to it," he recalled. They placed the fish inside the head coil and ran a series of images featuring human faces in social situations, "asking" the salmon to determine what emotion each person in a given photograph was experiencing. When he analyzed the data, he found a signal in response to the stimuli, indicating brain activity—in a dead fish.*

What could I hope to learn about myself from a brain scan? It depends on the scan. Were I to get a diagnostic medical scan, it might assure me that I don't have a tumor, or damage associated with a neurodegenerative disease like Parkinson's or Alzheimer's. Perhaps an individual fMRI—as opposed to the group study Eagleman's lab was conducting—might reveal certain personality traits, my emotional reactions to a given stimulus, my political orientation, whether I might have a drinking problem, whether I am lying or believe in God, what video clip I am watching on YouTube during a scan, or I might discover I am less empathetic when a Muslim hand is poked with a needle, as opposed to a Christian or Hindu hand. Brain scanning has been used to investigate these and many other questions, with varying degrees of success.

fMRI and similar cutting-edge imaging techniques are part

---

* Critics of fMRI studies often cite the dead salmon paper as evidence of the unreliability of the technique. It actually demonstrates the importance of correcting your statistical analysis to account for potential false positives.

of a long tradition of mapping the human brain. Even more so than our genes, the brain embodies the core of who we are. Dissection was the primary means for exploring the brain for centuries, until the invention of the microscope gave anatomists a closer look at brain tissue in the 1660s. Those early anatomical sketches were arguably the first crude brain maps. But how the brain actually works remained shrouded in mystery for two hundred years, until scientists began to link those anatomical maps with specific functions.

Form and Function

As a young medical student in Vienna in the late eighteenth century, Franz Joseph Gall envied his rivals' ability to memorize large amounts of information, far more than he could manage, despite the fact that he considered himself more intelligent. A humbler man might have considered the possibility that perhaps he wasn't as smart as he thought, but Gall nixed this explanation once he noticed that those students who proved most adept at memorization also had telltale bulging eyes and prominent foreheads. Clearly, he reasoned, the part of the brain used for verbal memory was located there, and was larger in those particularly skilled in that faculty. Exercising muscles made them bulge, so why shouldn't the muscles of the brain do the same? He decided that this bulging would be manifest in bumps around the head.

Once Gall completed his studies, he set about gathering further evidence to support his hypothesis that the brain was divided into dozens of "personality organs," and that the shape of the skull showed evidence of bony "bumps" that could be "read" by groping the subject's head to determine personality traits. He

scoured the city for working-class people in a wide range of oc-
cupations, and convinced them to participate in his research by
plying them with beer, wine, and occasionally cold hard cash. In
return, they told him all about their good and bad qualities, and
agreed to let Gall examine their skulls.

All told, Gall identified twenty-seven different telltale
bumps. There was an Organ of Theft, commonly found in pick-
pockets, an Organ of Murder, and (my personal favorite) an
Organ of Amativeness—a bulge at the base of the skull supposedly found in those with rapacious sexual appetites. Gall discovered it quite by chance when one of his subjects, a lusty widow
with quite the scandalous reputation, conveniently fainted in his
arms.

Gall called his new science cranioscopy, and he soon earned
fame and fawning acolytes across Europe. His protégé, Johann
Spurzheim, renamed the practice phrenology when he brought
it to the United States in 1832, where two brothers, Lorenzo and
Orson Fowler, took up the cause with a vengeance, even main-
taining a macabre phrenological cabinet featuring the plaster
casts of the skulls of murderers, madmen, and other fine specimens of humanity. The American poet Walt Whitman was a
lifelong fan of phrenology. Mark Twain was not, and was
hugely entertained when Lorenzo examined his head and pronounced it "generally mediocre" and "utterly lacking the Organ
of Humor." Twain's skepticism was well founded. Respectable
scientists had largely denounced phrenology as pseudoscience
by the end of the nineteenth century, although that didn't stop a
self-styled Australian phrenologist named A. S. Hamilton from
examining a wax "death mask" of the just-hanged outlaw Ned
Kelly in 1878. He concluded that Kelly's skull showed signs
of "dangerous over-development" in regions associated with

"combativeness and destructiveness," indicating an excess of self-esteem.

New York University neuroscientist David Poeppel laments the fact that Gall is best remembered for his work in phrenology, because Gall was also the first to propose a "parts list" for the human mind. The technical term is organology, in which different parts of the brain can be linked to specialized functions. Today we call this functional localization, and it is the essence of brain mapping. The rise of organology coincided with a number of unusual cases involving brain lesions, providing evidence that certain key functions, like speech, or vision, seemed to correlate with specific parts of the brain—and that damage to those areas of the brain could result in equally specific loss of function.

In 1861, a French anatomist and surgeon named Pierre Paul Broca encountered a thirty-one-year-old man named Leborgne, who had lost the ability to speak ten years before. Paralysis set into his right arm and leg, followed by gangrene, and he died in the hospital just six days after Broca first examined him. During his postmortem examination, Broca found that Leborgne's brain had a large lesion in the frontal lobe of the left hemisphere. Just a few months later, an eighty-four-year-old man named Lelong died after suffering a stroke that rendered him speechless: he could utter only five words. Broca examined Lelong's brain and found another lesion in roughly the same area as Leborgne's. Broca concluded that this region was the center for speech, now fittingly known as Broca's area.

Even so, there were hints that matters might be a bit more complicated. Back in September 1848, a young railroad worker named Phineas Gage was excavating rocks in Vermont to make way for a spanking-new railroad track. The process involved drilling holes in the rock and filling those holes with dynamite, using tamping irons—similar to crowbars—to tamp down gunpowder into the holes. But striking an iron rod against rock also produces the occasional spark. One such spark ignited the gunpowder: *kablooey!* The explosion drove the rod clean through Gage's brain, entering just under the left cheekbone and exiting through the top of his head. It landed some thirty yards away, "smeared with blood and brain."

Nobody expected Gage to live; his family had already begun making burial preparations. But despite developing an infection and falling briefly into a semi-coma, Gage not only survived, he was able to function surprisingly well within the year, "in full possession of his reason." Physicians who examined him at the time of his death eleven years later noted severe damage to his frontal lobe, which includes Broca's area, and yet Gage's speech was unaffected by his injury.

Any given area of the brain can be involved in many different functions, playing a slightly different role in each. The occipital lobe receives and processes visual information from the eyes, while the temporal lobe processes sounds detected by the ears. Nestled within the temporal lobe is the almond-shaped amygdala, which is associated with social and sexual behavior and other emotions, such as the fear response. Yet none of these regions operates independently of the others; they are all interconnected, forming circuits and sharing data to ensure we can adapt and survive in a constantly changing environment.

Even what we now call Broca's area is not precisely the same region identified by Broca himself as being crucial for the articulation of speech. In 2007, scientists at the VA North California Health System borrowed Leborgne's and Lelong's pickled brains and scanned them. The scan revealed that in both cases, the worst damage was in the region just in front of Broca's area, and it went deeper than just the surface of the frontal lobe, into the insula—typically associated with emotion, perception, motor control, and self-awareness—and the basal ganglia, located on either side of the thalamus, which have been linked to habitual behaviors.

For many years, neuroscientists puzzled over the role of the hippocampus, which serves as the seat of short-term memory. It didn't seem to do much of anything in countless brain-scanning studies. Then they realized that the hippocampus is involved in so many different tasks that once all the processing was done, the remaining signal wasn't strong enough to reach the critical threshold. The entire brain is engaged in any cognitive task. One 2012 study performed fMRI scans of 1,326 subjects all performing the same task, while another scanned just three subjects who repeated the same task five hundred times. The large data set allowed the researchers to detect effects that would otherwise be too small to meet the usual statistical threshold. Both studies found that the whole brain lit up, to varying degrees, rather than just selective areas, even when participants were performing very simple tasks, although the signal was stronger in some regions than in others.

A more accurate description of the brain's form and function would be a collection of interconnected neural networks, according to Ricardo Gil da Costa, a neuroscientist at the Salk Institute. "There is an overlapping of functions, shared regions," he said. "Today, we think in terms of different areas working together to perform specific cognitive tasks." That includes how the brain processes information about the self.

Accustomed to My Face

The man responsible for the first detailed, large-scale functional map of the human cerebral cortex was a neurosurgeon named Wilder Penfield. Penfield specialized in treating epileptic seizures, often resulting from head trauma like that suffered by Phineas Gage, who survived impalement by an iron rod only to be done in by violent seizures. Early in his career, Penfield collaborated with a German neurosurgeon named Otfrid Foerster who used electrical stimulation to operate on the brains of epileptics while the patient was conscious and under local anesthesia. The two men published the first cortical map in 1930, based on their observations during more than one hundred operations.

Penfield continued to refine the technique over the next twenty years, using his thriving neurosurgery practice to meticulously map out the brain, tying location to function. First he would apply a local anesthetic to the patient's scalp, removing a segment of bone to open up the skull and exposing the cerebral cortex. Much of the brain was still uncharted territory, so whenever he found himself operating on an unfamiliar area, Penfield would use platinum electrodes with nonconductive glass handles to apply a mild electrical current to the brain tissue in that region. Since the patient was conscious, he or she could describe any bodily response to that stimulation. He then marked that spot with a piece of paper to indicate which brain region elicited the response. Patients reported seeing stars when he shocked the visual cortex, or visual and auditory hallucinations—like hearing music playing—in response to probing of the temporal lobe. One woman memorably exclaimed, "I can smell burnt toast!"

This real-time feedback proved invaluable. Penfield was able to map out locations and responses to stimuli on postoperative sketches of his patients' brains, and noticed certain recurring patterns. Electrically stimulating a point just behind the ear produced a tingling sensation in the tongue; doing so just a little farther up did the same in the lips. In fact, as the shocks moved successively closer to the midline of the brain, patients reported sensations in the wrist, elbows, shoulder, trunk, hip, and knee, respectively. When Penfield applied an electrical current to the side of the motor cortex, he noted a similar sequence of responses to the stimulus, in this case involuntary movements in the jaw, lips, or tongue. Sometimes patients would involuntarily salivate. Moving progressively upward produced slight movement in fingers, then the hand, wrist, and elbow.

Ultimately Penfield operated on some four hundred patients and summarized his findings in a 1950 book, *The Cerebral Cortex of Man*, illustrated with drawings of grossly distorted human figures that we now know as the sensory and motor *homunculi* (literally "little men"). These constitute a visual representation of how the brain maps the body. The distortions in the homunculi result from the fact that, as far as the brain is concerned, the largest parts of the body are those that are the most sensitive, and hence the most interconnected. So in the drawings the hands, lips, and face are gigantic compared to the torso, for example.

These internal body maps are critical to how the brain distinguishes self from other. When we are just over one year old, we can look into a mirror and recognize the reflection as being "us," and the brain maps that visual input, tracking the changes in our size and appearance over the years. The brain also incorporates sensory feedback: if we knock on a door with our knuckles, we feel the impact, and this tactile feedback tells us the hand belongs to us, whereas watching someone else doing so doesn't elicit the same feedback (unless those knuckles connect sharply with our nose in a well-placed punch, in which case we recognize our broken nose as our own).

Both these processes involve a region in the back of the brain called the extrastriate body area that lights up during fMRI scans in response to looking at, or moving, a particular body part, including when we recognize a familiar face. It tracks other bodies, too, and uses its links to the sensory and motor cortices to determine what is self and what is other via sensory feedback. There is another region called the temporal parietal junction, which includes the angular gyrus. This gives you the sense of being in your body, and identifying where your body is located in space. It also seems to play a critical role in self-awareness.

One of the most active regions when it comes to our sense of self is the medial prefrontal cortex. It is part of what neuroscientists refer to as the default mode network: those regions of the brain that are most active when we are daydreaming, for instance, as opposed to engaged in a task that requires focused attention. Those areas show high baseline activity even when at rest, and that activity actually decreases when the brain is engaged in goal-directed activities that draw attention away from self-awareness. Whenever we lose ourselves in a given activity—what athletes sometimes call being "in the zone"—our default mode network is less engaged because we are less self-conscious.

Activity increases in the default mode network—and the medial prefrontal cortex in particular—in proportion to how relevant the information being processed is about ourselves, particularly when we observe our face or body in the mirror, or in a photograph. Jason Mitchell, a psychologist at Harvard University, conducted a study in which he videotaped participants during group activities and then showed them photographs of themselves or people they knew while monitoring their brain activity. The medial prefrontal cortex was more active when participants viewed images of themselves than when they looked at photographs of others, indicating that this region of the brain is critical to self-recognition. (Gil da Costa has observed the same effect in his work with macaque monkeys.)

This region is also where we store our representations of the people we know, and where we process social information, enabling us to predict how other people are likely to behave, and thus respond accordingly. "Humans are social creatures and the social world is a complex place," Cornell University cognitive neuroscientist Nathan Spreng told *Scientific American*. "A key aspect to navigating the social world is how we represent others."

Knowing where our body begins and ends and recognizing our face in the mirror are just the most fundamental aspects of how the brain creates our sense of self. All those interconnected systems within the brain also give rise to our temperaments, behaviors, emotions, memories, and beliefs. Yet more than 80 percent of all known human genes play some active role in the brain, and the brains of any two people are roughly 94 percent alike.

So what accounts for individual differences? New York University neuroscientist Joseph LeDoux has argued that it all comes down to the information encoded in the unique synaptic patterns in any one person's brain, which are determined partially by our genes, and partially by our environment. Understanding how neurons wire themselves requires taking a closer look at the microscopic structure of the brain and the means by which it forges all those crucial connections: through cells and synapses.

Butterflies of the Soul

In a hospital in Milan, a young physician named Camillo Golgi labored by candlelight in one of the kitchens that he had converted into a makeshift laboratory. By day, he treated patients. By night, he tinkered with chemical solutions, applying them to brain tissue and studying the stained tissues under a microscope, hoping to get a better look at their structure. One night, Golgi used potassium bichromate and ammonia to harden the tissue. Then he immersed the sample in a silver chromate solution. Usually, staining nervous tissue with other chemical compounds created a useless black blotch, but this time, Golgi found he could selectively stain silhouettes of nerve cells, which popped out against the translucent yellow background, clearly showing for the first time that a nerve cell had two types of projections: a long, slender fiber (the axon) and clusters of short, branching fibers (dendrites). This process became known as the "black reaction," and the discovery would revolutionize neuroscience.

The year was 1873. Two hundred years earlier, Robert Hooke had coined the word "cell" to describe the structure of a piece of cork he observed under his microscope. In the years since, scientists had determined that the tissues of most living things were comprised of cells. But the microscopes of the time weren't sufficiently powerful to make out much of the structure of brain cells, although scientists noted tangled fibers projecting from the cell bodies. Many subscribed to the theory that those jutting fibers fused together to form a seamless, intricate web of interconnected cells, much like the human circulatory system.

Then a young professor at Valencia named Santiago Ramón y Cajal made a fateful trip to Madrid to meet with Luis Simarro Lacabra, a psychiatrist who had just returned from Paris bearing brain-tissue specimens stained with Golgi's method. Cajal was writing a book on tissue-staining techniques and collecting illustrations to accompany the text. The specimens Lacabra showed him were a revelation. In his autobiography, Cajal described his reactions on seeing nerve cells "coloured brownish black even to their finest branchlets, standing out with unsurpassable clarity upon a transparent yellow background. All was sharp as a sketch with Chinese ink." He considered neurons to be the "butterflies of the soul."

Back home, he set up his own kitchen laboratory and improved Golgi's method, adding a second immersion to stain the tissue even more deeply. He drew lovingly detailed illustrations of the cell structure he observed under the microscope, including a famous depiction of the neural circuitry of a rodent's hippocampus. Cajal also applied the Golgi stain to tissues of the retina, the cerebellum, and the spinal cord. "As new facts appeared in my preparations, ideas boiled up and jostled each other in my mind. A fever for publication devoured me," he wrote. Cajal realized that all those long, slender cables jutting from cells remained distinct, without fusing into a mesh. He concluded that the nervous system is comprised of billions of separate neurons, communicating with one another via neural pathways that were called "synapses" for the first time in 1897.

Neurons make up the gray matter of the brain, what Agatha Christie's fictional Belgian detective, Hercule Poirot, called his "little grey cells." David Poeppel estimates that there are as many neurons in the human brain as there are stars in the Milky Way—around 100 billion.* There are many different types of

---

*In 2009, a Brazilian neuroscientist named Suzana Herculano-Houzel found that the human brain contains closer to 86 billion neurons.

neuron, depending on the function, but the basic structure has three parts: the cell body, the axon, and the dendrites. The axon is that long fiber Golgi observed sticking out of the cell. Axons make up the bulk of the brain's white matter, transmitting electrochemical signals between neurons across the many regions of the brain. The dendrites, or nerve endings, are smaller branches sticking out of the cell that allow the neuron to communicate with its neighboring neurons. A synapse is a small gap between neurons, across which electrochemical signals can travel from cell to cell.

The central nervous system forms a vast network in which cells that fire together, wire together.* While different regions of the brain are associated with different functions, and there are separate systems (such as Penfield's sensory and motor cortices), they are still part of a single brain. They might process different aspects of an experience—sights, smells, sounds—but it is still the same experience. Synaptic connections make this possible. When a group of cells in different interconnected regions fire simultaneously, they bind together, such that the next time a similar stimulus occurs, the same cells and connections will fire in response. Convergence zones—many found in the prefrontal cortex—integrate all that diverse information being processed in different regions of the brain to create a unified record of the experience.

This is the essence of plasticity: the notion that, while there is a fixed component of the brain (the hardware, if you will), the brain nonetheless "rewires" itself in response to experiences. Plasticity is further enhanced by chemical neurotransmitters,

---

*This is known as Hebb's Rule, after the Canadian psychologist Donald Hebb, who proposed it in 1949.

such as glutamate, gamma-aminobutyric acid (GABA), dopamine, and serotonin. The cells that produce each kind of chemical are generally found in the brain stem, but their axons snake into many different brain areas. These chemicals function as alarm systems, flooding synapses in response to specific stimuli. Only those synapses that are already active when the neurotransmitters arrive will be affected, and in this way the chemicals enhance the likelihood that the activated synapses that fire together, also bond and wire together.

Synapses are not passive storage devices. They are modified by experience. This, according to LeDoux, is the key to how the brain shapes its unique sense of self: you are your synapses. "People don't come preassembled but are glued together by life," he writes in *The Synaptic Self*. "And each time one of us is constructed, a different result occurs."

To date, most brain mapping efforts have centered on identifying which parts of the brain are affiliated with specific functions, or staining single neurons to track them in the mass of brain tissue. Ideally, neuroscientists would like to trace in exquisite detail each and every "wire" in the brain: the dendrites and axons that form the synaptic connections between neurons and thereby give shape to the self. All the cool kids call this the "connectome"—a snazzy digitized map, or circuit diagram, of LeDoux's synaptic self. If you are your synapses, you are also your connectome.

Wired and Inspired

It would be easy to mistake MIT's Sebastian Seung for a young high-tech entrepreneur, with his spiked black hair and casual-chic wardrobe of jeans, sneakers, and the occasional ironic

T-shirt. In reality, he is a crack computational neuroscientist who is helping transform neuroanatomy with automated systems capable of taking a sample of brain tissue and generating a complete circuit diagram. Seung started out in physics, working on artificial neural networks, and that research served as a natural segue into the ongoing effort to achieve a complete map of the connectome. He compares the connectome to the route maps found in the back of airline magazines: "Just replace each city with a neuron, and each route between cities by a connection between neurons."

The discovery of DNA's double-helix pattern in the 1950s was a major breakthrough in genetics, but yielded only general information about the functions of individual genes. It was only when the Human Genome Project mapped the entire gene sequence that scientists were able to start really exploring the power of genetics. Along those lines, the National Institutes of Health launched the Human Connectome Project in 2009, a five-year, $40 million initiative to trace the connectomes of more than 1,000 people. Neuroscientists would like to have the equivalent of a Google Maps for the brain that would enable them to zoom in for close-up views of specific neurons.

Dividing the brain into regions that correspond to various functions might help us interpret the symptoms of brain injuries, but it doesn't shed much light on how the different regions perform their functions, coordinate with each other, or how our brains shape who we are. "Perhaps minds differ because connectomes differ," Seung mused. "We are the product of our genetic inheritance and our lifetime experiences. Genes have influenced your connectome by guiding how your neurons wired together during the development of your brain.

Experiences have also modified your connectome, because connections are altered by the neural activity patterns that accompany experiences. Your connectome is where nature meets nurture."

While neuroscientists understand there is a vast, tangled network of single neurons connected by synapses, they haven't yet mapped out each connection separately. "We can never obtain satisfying answers if we consider regions as the elementary indivisible units of the brain," Seung told *Scientific American* in 2012. "An obvious solution is to understand a region by subdividing it into neurons and figuring out how the neurons work together to perform the region's function." More important, it might shed light on the mechanisms of certain brain disorders commonly attributed to "faulty wiring," such as autism or schizophrenia.

Unfortunately, just like mapping the human genome, this is a tough nut to crack. The neurons and their connections are not neatly laid out in easy-to-navigate grids; they are clumped together in a dense, tangled mess, like so many spaghetti strands, making it difficult and time-consuming to trace the myriad connections. Most of us have a hard enough time unraveling all the tangled power cords to the plethora of electronic devices now found in every household. It's so much harder to unravel the brain's circuitry, given that the human brain has as many as 100 trillion synapses. Winfried Denk of the Max Planck Institute in Heidelberg, Germany, has estimated that producing a wiring diagram for a single cortical column—a single unit of neurons in the cortex—would take about three billion years using conventional methods. The humble roundworm, *C. elegans*, has a paltry 302 neurons, yet it took researchers more than ten years to compile a complete connectome for the little worm, earning them a well-deserved Nobel Prize.*

Neuroscientists don't yet have a comprehensive model of how the worm's nervous system produces behavior, but Martin Chalfie of Columbia University has used the wiring diagram to identify which of its neural circuits were active when the worm wriggled backward after being poked in the head, and which were active when the worm wriggled forward when poked in the tail. "Without it we simply would not have known which cells were connected to which," he told *Scientific American*. In similar experiments, scientists have used the connectome to investigate how the worm responds to temperature, chemicals, and mechanical stimulation, as well as how they mate and lay eggs.

In 2012, UCLA scientists simulated the connectome of Phineas Gage and compared it to the brain structure of 110 healthy men between the ages of twenty-five and thirty-six—Gage's age from the time of his accident until he died eleven years later. Their model enabled them to better trace the trajectory of the infamous iron rod through Gage's brain, so they could simulate a lesion in those areas where they believed Gage's actual lesions had been. While Gage's speech was unaffected, his wife and other close friends noted radical changes in his personality and behavior. He became more impulsive, selfish, obstinate, unpredictable, prone to "the grossest profanity," and "devising many plans of future operation, which are no sooner arranged than they are abandoned in turn for others appearing more feasible." The UCLA model showed that the rod destroyed

---

*The 302 figure is for the hermaphrodite (female) of the species. Male nematodes have eighty-one extra neurons, mostly in the tail.

roughly 4 percent of the man's cerebral cortex and around 11 percent of the white matter in his frontal lobe, including the "hubs" in the frontal cortex that connect to other areas of the brain. Those injuries would certainly account for Gage's abrupt change in personality and behavior.

It could be argued that we already know all of this from the surviving historical documents. Not every neuroscientist is as bullish about the potential for the connectome to reveal the inner workings of the brain. "It's like a road map that tells you where cars can drive, but does not tell you when or where cars are actually driving," Columbia University's Oliver Hobert told *Scientific American*. Caltech's Christof Koch thinks that even such a detailed brain map would still be an oversimplification, noting, "Even though we have known the connectome of the nematode worm for twenty-five years, we are far from reading its mind."

One reason is that there are so many different types of neurons, and they are not interchangeable. Within the nematode's 302 neurons, there could be as many as 100 different types. Seung acknowledges this is far too many to explain the organism's behavior based solely on its connectome. When it comes to human beings, because the brain rewires itself constantly in response to experience, you would need many different connectomes for each individual to create a truly detailed map of synapses.

Mapping the human genome did not unlock all the secrets of our DNA, as previously hoped. Genes turned out to be far more complicated than scientists ever imagined; we have only begun to unravel the myriad factors that contribute to gene expression in individuals. At best, mapping the connectome will give neuroscientists a working approximation, which is arguably better than nothing. "This is just going to tell us where to look; then we need to study actual cells to learn more," Seung told *Nature*. "We know so little that a map of the connectome would be a useful place to start."

Cracking the Code

In the end, my brain scan at Eagleman's lab didn't tell me much about myself as a unique cogitating individual. That is largely because of the kind of study Eagleman was running: a group study versus an individual scan; the latter employs a different methodology and is much more difficult to interpret. "In one approach, you're getting a broad general sample and can pinpoint commonalities across populations," Gil da Costa explained. "In the other, you're potentially better able to pinpoint individual idiosyncrasies." You still need a statistical threshold in either case, but with the group scan, the objective is to look at variance across individuals within that group, rather than looking for variations across sessions involving one person. To compensate for the small sample size (N=1) in the latter case, a researcher must run multiple sessions with the same subject. I would have had to take that cognitive test many times to reach the required statistical threshold for a solid correlation.

"If I just know your height and you ask me to determine if you are male or female, I can't do that," explained Bradley Voytek, a neuroscientist at the University of California–San Francisco—even though height is highly heritable and strongly correlated with gender. "If you only give me one data point, I can't make that kind of prediction. It's the same with fMRI. [It] is a very powerful tool, but there are limitations. We have evolved to see patterns and we mistake those for causes very easily." Voytek draws an analogy to running on a treadmill while hooked up to a machine that records muscle activity from the arms and legs. The arms will move faster and faster as the speed of the treadmill increases, so there is a strong correlation between how fast the subject is moving and the rate at which the subject swings his or her arms. "That doesn't mean that the arms are where running happens, even though such a strong correlation can't be due to chance," he said. "If you don't have arms, you can run just fine."

It's important to bear this in mind, particularly when it comes to using fMRI as a means of lie detection. If a person is "dissembling," proponents maintain that it should be possible to tell that the person is lying just by looking at their brain scan, and U.S. courts seem to agree, allowing neuroimaging data to be submitted as evidence—at least for now. Melissa Littlefield of the University of Illinois has countered that this claim is based on fundamentally wrong assumptions, most notably the assumption that "truth" is the natural state of being—the baseline—and lying adds a false layer on top of the truth. An fMRI scan might reveal a lie if the person consciously knew he or she was lying. But what if someone doesn't realize they are lying, or has told the same lie for so long that it has become their truth? It's also possible to beat the machine: just move your head slightly, since fMRI requires the subject to hold perfectly still to get a usable image.

Another reason it is difficult to discern whether someone is lying on an fMRI is that human behavior is complicated, and thus far more difficult to "operationalize," according to Voytek. "If you can't operationalize your term biologically, then you can't make that determination scientifically with sufficient rigor," he said. Ultimately, it comes down to statistical significance. "If you put one hundred people in a scanner, I can probably tell you that a subgroup of forty-five people will probably lie more often than another subgroup of fifty-five people, but I can't tell you who was lying within their subgroups," said Voytek. "I have a statistically significant understanding of the group as a whole, but I can't say anything about an individual."

Even if we can't do lie detection, there *has* been impressive progress on single-subject fMRI, most notably by Jack Gallant at the University of California–Berkeley. Gallant studies the visual system. His subjects—in this case, two of Gallant's research team members—looked at thousands of pictures while the machine recorded the activity in their visual system. From this, Gallant built a model to predict their response to images they had never seen before. Gallant next picked 120 images not used in the first scan and used the prior results to predict the two subjects' brain responses to the new images.* Then the subjects were shown the 120 images and their actual brain responses were compared to the prediction. Using that model, Gallant and his team were able to reconstruct what the subjects were seeing—or at least a fuzzy, crude semblance thereof—with 90 percent accuracy. Gallant has compared this to a magician asking a mark to pick a card, any card, without resorting to psychological tricks, using a computer algorithm instead.

Three years later, Gallant was back in the news for using a similar fMRI-based model to reconstruct moving images as

---

* Sample size is key. Gallant found that the statistical tipping point for the model's accurate predictions was 120 images in the set.

viewed by his test subjects. Subjects watched several hours of movie previews while in the scanner, and this data was deconstructed to identify the activation patterns for each second of video footage. From this, Gallant's team built another model linking "the plumbing to the blood flow that you do see with fMRI to the neuronal activity that you don't see," Gallant told *Technology Review*. Then they assembled a library of 18 million video clips from YouTube, randomly chosen. That library was used to simulate the likely readings of the fMRI images as subjects watched a new set of movie trailers. The result: the simulations and the real-time scans were nearly identical. It isn't technically mind-reading, but it's the closest neuroscience has come so far.

For all the progress made in brain imaging and mapping to date, the philosophically minded David Poeppel questions the wisdom of continuing to focus so heavily on what he has termed the "cartographic imperative"—the need to tie neural functions to specific locations via a progressively more detailed brain map. He believes this is unlikely to provide a satisfactory explanation for deeper questions in neuroscience—namely, the neurological basis of human perception, the acquisition of language, and the Ultimate Question underlying it all: human consciousness. He doesn't deny the paradigm has been valuable; mapping form with function certainly shed light on the brain's sensory systems (vision, hearing, body maps). But it has been less helpful at illuminating more advanced cognitive functions, such as memory.

An fMRI reveals how specific tasks correlate to different parts of the brain by measuring blood flow; it does not directly measure neurons firing, and it cannot reveal how many neurons are firing, or how those neurons affect other brain regions. "An image by itself, just as a word by itself, is not enough to provide the infrastructure for explanation," Poeppel argued in a 2008 paper. "The fact is that there can be no objective image of the mind at work. Imaging generates data, no more, no less, and data are only useful and usable insofar as they have an interpretation in the context of a theory about how the mind is organized."

Cardiff University's Dean Burnett has compared scanning technology in neuroscience to the discovery of the Valley of Kings in Egyptology. "We now have access to vast amounts of previously inaccessible information," he observed in the *Guardian*. "But just because you can see the hieroglyphics, it doesn't follow that we know what they mean." Neuroscientists are still searching for their Rosetta Stone to help them crack the code.

3

Moveable Types

And then I was being chased by an improperly filled in answer bubble screaming, "None of the above!"
—BUFFY SUMMERS, "BRAIN CANDY"

When I was in high school, my mother became fascinated with a book by a minister named Tim LaHaye, *Spirit Controlled Temperament*, which divided people into four basic personality types: sanguine, choleric, phlegmatic, and melancholic. Since hardly anyone fit perfectly into just one category, there were twelve pairs of hybrid subtypes for good measure. My mother enthusiastically sorted our entire family into these subtypes, based on the checklist descriptions. She decided that my highly sociable, somewhat blustery and aggressive father was totally a "Chlor-San"—a combination of choleric and sanguine—while she was a "San-Phleg" and my sister was a "Phleg-San," both having the same social and fun-loving instincts as my father, tempered by the relaxed warmth of the phlegmatic. My brother she deemed a "Phleg-Chlor"—basically easygoing with fiery accents. With my bookish, dreamy, occasionally anxious and moody nature, I clearly had to be a "Mel-Phleg," the sole melancholic in our suburban household.

LaHaye was not the first to link personality to these four categories. His "spiritual temperaments" are a crude reworking of the theory of the four humors, dating back to Greek and Roman antiquity, when leading physicians and philosophers used it to explain ailments in both mind and body. They believed that the body contains four basic fluids, or "humors," which must be in balance in order for a person to be healthy: black bile, yellow bile, phlegm, and blood. It's unclear how the notion developed, but in 1921, a Swedish physician studying how blood clotted noted that if one drew blood and let it sit for a few hours in a clear glass container, it would separate into four distinct layers. There would be a blackish sludge at the bottom (black bile), then a layer of red blood cells, followed by a layer of white blood cells (phlegm), topped off with a clear yellow fluid (yellow bile). It seems as good an explanation as any for how this became the centerpiece of ancient medicine.

The balance of humors could be disrupted by eating the wrong kinds of foods, or by inhaling mysterious "vapors," thereby resulting in any number of maladies. Too much yellow bile? You're choleric, prone to angry, aggressive behavior. Too much phlegm? You're afflicted with a certain lassitude, or possibly epilepsy.* An excess of blood? That would make you sanguine, overly excitable with a flushed, ruddy complexion. An excess of black bile would make you melancholic. Those hu-

---

* Galen believed that epilepsy was due to excessive phlegm up the nose. He advised a tincture of mistletoe, peony, and powdered human skull as treatment.

mors corresponded to four basic temperaments, the same ones handily co-opted by LaHaye.

If you're not keen on medieval humors for insight into your nature, there is always blood type, one of the most popular methods of categorizing personality in Asia, where it is known as *ketsueki-gata*. In many Asian countries, you are what you bleed. Women's magazines in Japan have blood-type horoscopes, combining blood type with both Western and Chinese zodiac signs to create elaborate "love biorhythm" charts. Japanese employers have been known to discriminate against job applicants with Type B blood, believing them to be flighty and unreliable. There isn't a shred of scientific evidence to support this: study after study has found no correlation between blood type and personality factors. But lack of empirical evidence never dissuaded a true believer.

Or perhaps you are what you sweat. A 2011 study had a group of men and women sleep in the same white T-shirt for three consecutive nights before sealing it in a plastic bag. Then the researchers brought in a team of two hundred "sniffers" to take a whiff of each shirt and guess how anxious, dominant, or extroverted the wearer might be. While none of the sniffers should expect a call from the Psychic Friends Network anytime soon, they were just as accurate in their personality assessments as the control group, who based their assessments on viewing videos of the subjects. The oddball research has already found its way into so-called pheromone parties for singles, at which guests fork over an admission fee for the privilege of sniffing malodorous T-shirts in handy Ziploc bags in hopes of finding that special someone through their unique scent.

What is this mysterious quality we call personality? Many

psychologists would define it as enduring patterns of thoughts, feelings, and behaviors that distinguish individuals from one another—that is, the sum total of the mental, emotional, and social characteristics, including quirky behaviors, that make us who we are. Psychologists call these characteristics traits, and they are the fundamental units of personality. Just as one person has blue eyes and another brown, so too is one person more anxious, outgoing, or detail-oriented than another. It's why people respond quite differently to the same situations and stimuli, and it is the source of more than a little social conflict.

There are two kinds of personality traits: those related to temperament and those related to character. Your temperament is largely due to your genes and biological factors outside your control. Your character, on the other hand, stems from your home environment, relationships, and personal experiences, and is shaped further by countless cultural factors. "Your character traits stem from your experiences," psychologist Helen Fisher wrote in *This Will Make You Smarter.* "The balance of your personality is your temperament, all the biologically based tendencies that contribute to your consistent patterns of feeling, thinking and behaving." Temperament is largely fixed; character is flexible, enabling us to adapt and evolve to changing circumstances as we go through life.

Combined, these traits make up your personality and manifest in your behavior. It sounds so neat and tidy, but unlike measuring height or weight, it's impossible to observe a personality trait directly. Most assessments of people's traits are based on observations of their behavior—and when it comes to behavior, human beings are complicated and unpredictable. Psychologists have struggled for more than a century to devise rigorous and

robust methodologies to extract meaningful correlations between behaviors and personality traits from the raw data.

The human compulsion to categorize our personalities into
neatly labeled boxes goes back to the dawn of astrology, quite
possibly the oldest personality typing system in history. Have
you ever taken a Facebook quiz to determine which superhero
most resembles you, or completed a compatibility quiz for an
online dating site? You are not alone. The enormous popularity
of such instruments stems from the same pattern-seeking compulsion. It is one of the many ways we draw a boundary between self and other, defining who we are by identifying how we
are different from those around us.

Personality assessment is also a lucrative business, and an
active area of psychological research. There is a vast spectrum of
questionnaires, inkblots, and other psychological tools in use all
over the world. The plethora of testing instruments prompted
psychology pioneer Gordon Allport to lament in 1958, "Each
assessor has his own pet units and uses a pet battery of diagnostic devices," making it difficult to compare different psychological studies of personality. Most of us will encounter a personality
test at some point in our lives. By 2003, personality testing was a
$400 million industry, with many Fortune 100 companies regu-
larly administering the best-known personality test to their em-
ployees: the Myers-Briggs Type Indicator (MBTI).

Hidden Dimensions

There is nothing quite like the pain of a failed relationship to
spur one toward renewed introspection. Such was the case with
famed psychologist Carl Jung, who found himself emotionally

shattered by a falling-out with his friend and mentor, Sigmund Freud. The two men met in 1907, after a starry-eyed Jung, then thirty-one, sent the famed psychoanalyst a copy of a volume he'd supervised, *Studies in Word-Association*, which fervently praised Freud's pioneering work. Both felt an instantaneous connection. Their first conversation lasted thirteen hours, and ignited an intellectual "bromance" fueled over the next six years by the many letters they wrote to each other.

Their split is usually attributed to intellectual and philosophical differences about the young field of psychoanalysis, and those differences were indeed substantial. But the conflicts were exacerbated by tensions stemming from their equally contrasting personalities. Their correspondence, first published in the 1970s, sheds light on the unusual dynamics in the relationship. Freud was the powerful father figure and reveled in Jung's youthful enthusiasm, seeing the younger man first as his intellectual soul mate and heir apparent (he once addressed Jung as "spirit of my spirit"), and later, when the friendship soured, as a rival and potential usurper. "He will grow, and I must dwindle," Freud lamented in a letter to Jung's wife, Emma. Jung, for his part, delighted in the attention and affection of such an illustrious mentor, but grew to resent the "father creator," complaining of having to live off "the crumbs that fall from the rich man's table." Jung hinted at increasing discomfort with Freud's constant need for assurances of devotion and loyalty, alluding in one letter to an incident of sexual assault by a previous mentor. By 1912, their once-strong bond was broken.

The disintegration of his relationship with Freud left Jung in a deep depression, and he found himself pondering their pronounced differences in temperament. He concluded that Freud was fundamentally an extrovert, energized by frequent social contact, while Jung was an introvert, energized by solitude. These two kinds of orientation—outward and inward—formed the first of Jung's core dimensions of human personality. He presented his nascent theory the following year at yet another conference in Munich, coincidentally the last time he would ever see Freud.

Then Jung recalled a quarrel with his wife, Emma, over his behavior at a dinner party. She accused him of being rude and tactless, leaving several guests with hurt feelings. Clearly, he reasoned, there were "feeling" types, like Emma, and "thinking" types, like the cool, calmly aloof (and no doubt greatly misunderstood) Jung himself. This gave him a second two-pronged personality dimension, one describing "rational" or "judging" functions. To round things out, Jung hit upon a third dimension: intuitive and sensing, two different approaches to what he deemed an "irrational" or "perceiving" function. These three dimensions became the basis of Jung's seminal treatise, *Psychological Types*, which was translated into English in 1923, and promptly came to the attention of an intellectually curious housewife in Washington, D.C., named Katharine Briggs.

Prior to her marriage, Katharine had been one of very few women in her era to receive a college education, enrolling at Michigan State University at the tender age of fourteen. She and her husband, the physicist Lyman Briggs, doted on their only child, Isabel, who proved to be as precocious as her mother, publishing her first short story at fourteen. Two years into her own college education, Isabel brought home a fiancé, Clarence "Chief" Myers. While Mr. and Mrs. Briggs approved of their daughter's choice, Katharine became fascinated by how markedly different in temperament her new son-in-law was from the Briggs clan. Chief was pragmatic, logical and detail-oriented, in contrast to his imaginative young bride. When Katharine read Jung's treatise, she latched onto his three dimensional personality types to explain the familial differences, just as my own mother did with LaHaye's four spiritual temperaments. All four of them were introverts, she concluded, but the men were thinkers and the women were feelers, while Chief was a senser and the Briggses were intuitives. Isabel, too, became enthralled by Jung's ideas, weaving them into her 1929 bestselling novel, *Murder Yet to Come*.

It might have ended there, had Isabel not stumbled across a 1942 article in *Reader's Digest* on the "people sorting instruments" popularized in the 1920s, rudimentary forms of personality tests used to "place the worker in the proper niche." Isabel eagerly investigated the effectiveness of the tests, convinced this could be a useful tool for women who were joining the industrial workforce for the first time during World War II. But she was disappointed to discover they were quite poor predictors of employee performance. Katharine suggested she make up her own test, building upon Jung's types, and Isabel pounced on the idea.

Isabel boned up on statistics and psychometrics, and even apprenticed herself to the Philadelphia-based founder of what would become a successful human resources consulting firm. She ended up adding a fourth dimension to Jung's original three: judging versus perceiving, which pertains to how we function in the world. Those with a judging bent are decisive, highly structured, and like to plan ahead, while those with a perceiving inclination are more spontaneous and adaptive, preferring to keep their options open as long as possible.

With all that research under her belt, Isabel devised a rudimentary prototype quiz, based on a "forced choice" format: the subjects responded to a series of descriptive sentences with a simple Yes or No answer. Those answers, in turn, were coded and compiled, and Isabel would assign the subject a four-letter personality type out of a possible sixteen. ENFJ, for instance, would describe someone who tested as extroverted, intuitive, feeling, and judging; while ISTP would describe someone who tested as introverted, sensing, thinking, and perceiving. Everyone possesses both sides of all four dimensions, she argued, but each person expresses a marked preference for one pole or the other in each dimension. To prove the robustness of her model, Isabel needed data—a lot of data. So she began giving the prototype quiz to family, friends, and anyone else she could persuade to participate, obsessively revamping her prototype test as more and more data came in.

It took well over a decade for Isabel's peculiar hobby to be accepted as a legitimate instrument, but the Educational Testing Service published the first complete Myers-Briggs Type Indicator in 1962. It has withstood the test of time; more than two million people take the MBTI each year. So the MBTI seemed the obvious choice to assess my own personality quirks. I forked over the fee, answered the seventy-odd forced-choice declarative statements, and soon found myself on a conference call with one of the testing company's staff counselors, a charming and amiable man named Alan, who walked me through my results.

His first question was deceptively simple: "Look out the window and describe that."

"Uh—describe what, exactly?"

Alan stuck to his guns. "Look out the window and describe that."

"You mean the scenery?"

"Look out the window and describe that."

I finally described the scenery outside my office window, just so we could move forward. Alan confessed that the wording was designed to elicit just such a reaction, at least if one falls onto the intuitive side of the intuitive versus sensing dimension, which describes how we gather data and perceive the world. Most people—70 percent of those who take the test—are sensers, and describe what they see out the window, no questions asked. Alan explained that while sensers tend to be literal-minded and detail oriented, fuzzy-headed intuitives like me immediately start imagining the different possible meanings inherent in the question. We don't just collect data, we look for patterns, and frankly, we can be a bit sloppy about the details, which drives the sensers crazy. I scored on the moderate side of the intuitive scale, because it's not like I have no detail-oriented skills whatsoever. But my natural preference falls more on the intuitive side.

Alan next presented me with an imaginary dilemma: I am involved with a local youth sports league, and my team qualifies for the state championship. But there is not enough money in the budget to take the whole team. What solution would I suggest? That seemed straightforward enough: I said I would figure out how much we were short and throw a fund-raiser to try to make up the difference. If that failed, we would regretfully take the best players to the championship to ensure we had the best chance of winning. This makes me a thinker as opposed to a feeler, even though I emphasized that I would feel really bad about it. (I scored near the bubble in that category.) Alan swore that leaving any player behind would be unthinkable for true feelers, even if they had to make up the difference out of their own pockets—those old softies—whereas thinkers wouldn't see the point of sacrificing the team's chance at a championship just to spare the weaker players' feelings. We would bulldoze over any personal objections for the greater good of the team.

This is the only category where one sees a strong division along gender lines, according to Alan, with two-thirds of women testing as feelers and two-thirds of men testing as thinkers—a breakdown that seems to be at least partially due to cultural conditioning. "Strong, objective, analytical women often find themselves in positions of leadership, but these qualities are not seen as positive in women," Alan explained. "There's another word we use to describe such women." I have been called that word many times in my life. Those poor imaginary teammates who didn't get to go to their state championship would most definitely have hurled that epithet at me. And I would have told the brats to suck it up and learn to deal with life's disappointments.

Harsh? Perhaps. But it doesn't really make me a stone-cold bitch, any more than a feeler is just a sappy pushover. The dimension simply describes one's approach to decision making: do we focus on the people, or on the problem? I prefer to work the problem, although ideally, we should focus on both. Too much emphasis on not hurting people's feelings clouds one's thinking and makes it difficult to find the optimal solution. Yet not taking people's feelings into account at all rather defeats the purpose of problem solving in the first place: to make things better for everyone. "The best decisions are made by marrying the two ways of thinking," Alan advised.

I scored highest on judging (as opposed to perceiving), at the top end of the moderate range: I'm definitely a planner, a maker of lists, a decider, and occasionally a bit of a dictator if it looks like a situation is getting too chaotic, although I've become far more flexible with age. The most surprising result was my score on the extroverted/introverted dimension: I scored (just barely) as an extrovert. That is simply not correct, even though it was based on my self-reported answers. I suspect the wording of those questions was sufficiently ambiguous that, given the forced-choice nature of the test, my answers proved misleading. I am a classic introvert, although I have become more extroverted over the years because my professional life demands it. While I enjoy socializing, I prefer small dinner soirees to loud, raucous parties, and need a great deal of time alone to recharge my social batteries. I guess that makes me an extroverted introvert.

In the end, Alan opted to override that score, and I came out as an INTJ: introverted, intuitive, thinking, and judging. When I first took the MBTI in my twenties, I tested as an INFJ; over twenty years, I switched from feeling to thinking, and also became more extroverted. Apparently this isn't unusual. Less than half (47 percent) of those who take the MBTI get the same type when they retake the test, even if they do so just a short while later. A 1991 report by the National Research Council estimated that between 24 percent to 61 percent of those who take the MBTI will get the same type if they retake the test within five weeks to six years—which means that 39 percent to 76 percent would get a different type. The NRC report wisely recommended caution in making major life decisions based on one's MBTI category. (So did Alan, for that matter.)

Yet the MBTI insists that one's core personality type doesn't change: once an INFJ, always an INFJ. The NRC findings are explained away by asserting that it is possible to develop skills in those areas that are not one's natural personality preference, much like we are born either right- or left-handed, but can learn with practice to use our non-preferred hand. So which of my test results was wrong: the INFJ in my twenties, or the INTJ in my forties? And which represented my "true" preferences? This insistence on the permanence of personality type is a departure from Jung, who believed that personality was not static and immutable but continued to evolve over the course of one's life. "Every individual is an exception to the rule," Jung opined, deeming any attempt to force people into a rigid system "futile." He also dismissed easy labeling of people based on limited observation as "nothing but a childish parlor game." Ouch.

Regardless of whether I am an INTJ or an INFJ, I am a very special snowflake: only 2 percent to 3 percent of the population falls into each of those two categories.

The harshest critics of the MBTI dismiss it as little more than a "Jungian horoscope," with the sixteen types having no more validity than the twelve signs of the zodiac—or the four medieval humors. Back in 1949, a psychologist named Bertram Forer conducted a study in which participants were given short written descriptions of their personalities based on their astrological signs and asked to rate the accuracy of those descriptions on a scale of 1 to 5. Actually, Forer randomly distributed different glowing descriptions of various star signs gleaned from an astrology book, and it didn't matter whether the description pertained to the subjects' actual star signs or not. The average rating for accuracy was a whopping 4.2, with more than 40 percent giving their descriptions a perfect score. This is known as the Forer effect.

I had my astrological chart done in my youth just like any other hip, young, would-be Bohemian. I had always hated my sun sign, Taurus. The descriptions invariably included words like "practical," "stubborn," "deliberate," "reliable," or "lazy," not to mention being obsessed with material things and prone to overindulge in food and drink, thereby packing extra pounds on our stocky frames. That's right, fellow Taureans: we are the dreary, overweight tax accountants of the zodiac. This didn't fit my view of myself at all; I longed to be something more colorful and carefree, like a Gemini or a Leo. So I was delighted when my full astrological chart revealed that my moon is in Leo, with the other planetary alignments and their placement in specific "houses" at the time of my birth, supposedly conferring a whole host of desirable personality quirks upon me. It provided a far more complex and flattering self-portrait than my plodding sun sign alone, especially when combined with my sign in the Chinese zodiac (Year of the Dragon). How could it not be accurate?

That is the Forer effect in action. Those with a skeptical bent, like author Annie Murphy Paul, believe it also explains the appeal of the MBTI, which is relentlessly positive in its descriptions of the sixteen personality types, despite being draped in the sheen of science. "People are more likely to endorse positive accounts of themselves," she writes in *The Cult of Personality Testing*. Paul thinks the MBTI is so seductive precisely because of "its reassuring confirmation of what we already know about ourselves," and dismisses it as little more than a clever repackaging of "a shallow self-appraisal."

She has a point; the MBTI descriptions of the sixteen types are maddeningly vague. As much as I enjoyed my chat with Alan, I'm not sure how helpful it was to learn that I am "attentive to the inner world of possibilities, ideas, and symbols," with "deep interests in creative expression." But maybe that's just my inner thinker talking. It's certainly fair to say that the MBTI is an imperfect instrument; its predictions of type are only as accurate as the self-assessments of the test-takers, which are subject to the whims of the moment, and one's degree of self-knowledge. But people do find value in it, even if they only gain the realization that most personal conflicts stem from fundamental differences in temperamental styles. Any test that prompts people to step outside their bubble and consider another point of view has its uses. Problems arise when it is used as a basis for hiring decisions, to cite just one example of potential abuse.

However personally fulfilling the MBTI might be for diehard acolytes, forcing people into discrete categories isn't helpful to professional psychologists who study personality. "It puts people into a typology and it just doesn't work that way statistically," said Josh Jackson, a psychologist at Washington University in St. Louis (WUSTL), who admitted he views the MBTI as something of an embarrassing problem child. "People aren't this way or that. It's more of a gradation." The confusion over whether I was an extrovert or introvert, a thinker or a feeler, is a case in point. I was close to the cutoff point on the scale for both those dimensions. Answer just one or two questions differently, and I could easily find myself scoring as a different type.

Personality is a continuum. Newcastle University's Daniel Nettle argues that we all share the same basic traits, just as we all have a height and a weight; the only differences are the degrees to which we express those traits. So instead of placing me in a discrete fixed category for my personality "type," what is needed is a personality model that lets me know where I fit on the broad spectrum of a trait like extroversion, relative to the general population. Psychologists affectionately dubbed that model the Big Five.

Measure for Measure

Like many scientists, Francis Galton was accustomed to giving lectures. He noticed that one way to tell whether an audience was bored was to monitor the frequency of their fidgeting. He found that people fidgeted roughly once a minute on average during a lecture. If they were especially enthralled with the speaker, that rate was cut in half, and the fidgets were of shorter duration. But when the audience became bored, they fidgeted more often, and the fidgeting lasted longer. Galton also noticed that bored audience members slouched more in their seats. He even measured the precise angle of deviation from an upright sitting position, thereby giving "numerical expression to the amount of boredom expressed by the audience generally during a reading of any particular memoir."

Galton had a gift for measurement, amassing data on head size, the weights of livestock, and the shapes of fingerprints, and insisting that emotional responses could be measured physiologically through changes in heart rate. Speculating about personality traits was all very well and good, he reasoned, but ultimately you needed a good metric, and he found one in a dictionary. By Galton's estimation, there were at least 1,000 terms describing people's characters in the English language. It was just a passing observation, but it inspired the "lexical hypothesis," which asserts that the most important personality traits among people will, over time, become incorporated into language. Therefore it should be possible to sample language to identify descriptive adjectives and other words associated with those traits.

The problem is that language offers a pretty large data set. In 1936, Gordon Allport and H. S. Odbert combed through two enormous dictionaries, from which they identified nearly 18,000 words describing personality. (I'm guessing they would score as detail-oriented sensers on the MTBI.) They eventually settled on 4,504 adjectives to describe the most "permanent" personality traits. That is still a daunting number of variables. How do you winnow down such a substantial list even further? Psychologists found the answer in a statistical technique called factor analysis, which is designed to pinpoint underlying elements, or factors, that describe key differences across a large sample population. Jackson likens the process to sorting through a large pile of dirty laundry, separating the various clothing items into colors, whites, and delicates instead of trying to describe each item individually in great detail.

It's not as easy as it sounds, since there is no sample list of categories already in place; those categories must emerge from the raw data, usually collected via questionnaires asking people to rate the degree to which various descriptive sentences accurately describe their personalities. The subsequent analysis looks for correlation coefficients. For instance, there is a correlation between someone's height and weight of around 0.68; a perfect correlation would be 1, meaning that the variance in height perfectly predicts variance in weight. Height is just one factor that determines weight, so the correlation coefficient is not precisely equal to 1. But it isn't 0, either; a correlation of 0.68 is actually pretty significant. Factor analysis does the same thing for linking certain behaviors with specific traits, taking into account the correlation coefficients of all the variables involved. The more variables there are, the more difficult it becomes, and the easier it is to mistake coincidence for correlation.

Over the years, psychologists couldn't help but notice that the same five factors kept emerging from the raw data:

Openness, Conscientiousness, Extroversion, Agreeableness, and Neuroticism, handily summed up by the acronym OCEAN. These are known as the Big Five, currently the standard in psychology for conducting personality research. Openness is a measure of one's curiosity and willingness to explore new territory; those who score lower on this trait tend to be more traditional, preferring steady, predictable routine and sticking with familiar places and interests. Conscientiousness is a measure of how one controls and directs one's impulses, and a high score on this factor indicates someone who is disciplined, reliable, achievement-oriented, and probably a bit of a neat freak.

Extroversion measures the degree to which someone enjoys being with, and is energized by, socializing with other people. It is also marked by certain telltale behaviors, such as novelty-seeking and high activity levels. This is one of the most commonly misunderstood personality traits; most people assume it means being very outgoing and talkative, a "life of the party" type, when in fact it describes how people respond to stress and negative events, and what they need in order to recharge their emotional batteries. After a bad breakup, would you rather curl up with a pint of Häagen-Dazs and a pile of DVDs, or go out dancing with a group of friends? If the former, you're an introvert; if the latter, you're an extrovert.

The degree to which someone is compassionate, cooperative, and desires social harmony is defined by Agreeableness. People who score low on this scale tend to be more suspicious and non-trusting, even antagonistic—like those folks on certain reality TV shows who pride themselves on speaking the "raw truth" and loudly declare they're "not here to make friends." Finally, there is Neuroticism, the "downer" dimension of the Big Five. Again, people often confuse this factor with the common usage of "neurotic," usually employed to describe someone who is obsessive-compulsive, highly organized, or a chronic neat freak. But in the Big Five, these qualities would instead be indicators of high conscientiousness. A high score in Neuroticism describes someone who is prone to negative emotions and mood swings in response to stressful situations, and more likely to suffer from depression or anxiety.

Like the MBTI, a typical Big Five questionnaire consists of a series of statements describing personal preferences, but instead of the standard forced-choice yes-or-no format, participants have five response options for each question: very accurate, accurate, neither accurate nor inaccurate, inaccurate, or very inaccurate. The number of questions can vary widely, from more than one hundred to as few ten, as with Nettle's own Newcastle Personality Assessor (NPA), which provides just two questions relating to each of the five factors. Can ten questions really target someone's personality as well as one hundred? On the NPA, I scored in the low-medium range for Extroversion and Neuroticism, high on Conscientiousness and Openness, and high-medium on Agreeableness. Those were mostly in the ballpark when compared with my results on the longer version, except I scored slightly higher on Agreeableness and slightly lower on Neuroticism.

Several population-wide studies have demonstrated that people become more agreeable and conscientious as they age, better at inhibiting impulses, and also less anxious. But Jackson emphasized that this is not tantamount to changing one's core temperament. I have become more extroverted, less neurotic, and more open over the years, but everyone else has followed a similar developmental path. So I would still be roughly the same in relation to the population at large.

Just because five factors are the most common number of traits to emerge from the data, that doesn't mean that these are the only possible characteristics to explain the differences among people. "You can have five factors, or four, or sixteen, or even eighteen thousand adjectives," said Jackson. It all depends on how much detail you need. The five factors provide a broad overview, but it's possible to add extra factors one by one to gain a much more nuanced view, much like increasing the resolution of a microscope. So the Big Five would apply to the scale we can see with the naked eye, while the full ten thousand factors from which those five factors were derived would be analogous to the subatomic scale.

Columbia University psychologist Walter Mischel has argued that no such test could predict human behavior with any significant degree of accuracy. As evidence, he pointed out that the typical correlation between someone's score on a personality test and their actual behavior is only around 0.30, or less than 10 percent of the total behavioral variance observed in populations. According to Mischel, this is because our behavior is strongly influenced by the situation in which we find ourselves. In other words, we adjust our behavior to suit whatever social role we play in a specific context, such as the home or work environment. The current consensus among psychologists strikes a happy medium: behavior is the result of an interaction between core personality traits and contextual situations. Both variables must be accounted for, and while it's not possible to predict exactly how one person will react in a specific context, it is nonetheless feasible to identify recurring patterns of behavior over time.

The Big Five still relies on self-reporting, and it is fair to ask whether this approach truly produces an accurate assessment of one's personality. The answer might surprise you: it does, at least for certain aspects, and the accuracy of self-reports can be honed even further with feedback from one's close friends and family. Jackson's WUSTL colleague, Simine Vazire, has developed what she calls the self-other knowledge asymmetry (SOKA) model, which matches self-reports with third-party assessments of the subject's personality by those who know the subject well. This can be combined with more objective tracking methods, such as wearing an audio recorder for six days. Vazire then compares the tracking data and the self-reports with the assessments of friends and looks for areas of overlap and divergence. Giving the test multiple times can further weed out muddling effects due to error or randomness.

It turns out that people are surprisingly good at assessing their own personality traits, at least when it comes to their internal thoughts and feelings, like insecurity or anxiety. However, they are less adept at accurately assessing more external traits, like attractiveness, intelligence and creativity, or overt behavioral tendencies. Our friends, and even strangers, provide better assessments on that front. Furthermore, the feedback we receive from others can influence our self-perception, through subtle cues or reactions to our words or behavior. This works both ways: how we see ourselves can influence the perceptions of those around us. Merely admitting to friends that we were up all night worrying could make them view us as being more anxious than they did before. In fact, "Even if we never communicated, we probably have similar impressions," said Vazire, because we have access to the same informational cues. "It's important not to assume that all the agreement is because of communication and feedback. Some of it just because we are both perceptive people and we see the world as it is."

It is rare for there to be a substantial gap between your self-perception and how others perceive you; in fact, it's usually only seen in those with mental imbalances. In *Addams Family Values*, the voluptuous new nanny, Debbie Jelinsky, is really a serial killer known as the Black Widow, with her sights set on marrying Uncle Fester and then bumping him off to inherit his fortune. The first part of her diabolical plan is successful: the lonely Fester is an easy mark for her wiles, and soon finds himself living in shame in the suburbs, surrounded by gaudy architecture and pastels. But she discovers it's not that easy to kill an Addams: he survives numerous attempts on his life with oblivious good humor.

Her frustration forces a confrontation with the clan, whereby Debbie ties them to electric chairs in the basement and gives a PowerPoint presentation on her life history to illustrate why she's really the victim here. We learn that her homicidal tendencies emerged early, at the age of ten, when she requested a Ballerina Barbie for Christmas. When the day arrived, and she unwrapped the gift, what did she find? Malibu Barbie! "That wasn't what I asked for. That wasn't who I was," she rants to the bemused family members. "I was a ballerina! Graceful! Delicate! They had to go."

Debbie is enraged not by the gift, but by the fact that her parents do not see her as she sees herself. When an extreme gap occurs, it is probably because we have so internalized our own self-perception that we have become blind to how our actions are perceived by others. We forget that others can't read our minds and thus give us the same benefit of the doubt we so generously confer upon ourselves. Debbie sees herself as someone who just wants love, designer clothing, jewelry, and a Mercedes (or two), and feels victimized when people tell her she can't have her way. Her self-image is markedly different from how her victims see her—as a spoiled, greedy, manipulative psychopath with no moral qualms about attempting to electrocute an entire family. Debbie is blind to the reality of her behavior because she is solely focused on her own internal thoughts and needs.

Most perceptual gaps are not as severe, but we have all experienced that jolt of shock upon realizing that our view of ourselves does not necessarily match how others see us. Once, while leaving a matinee showing of *The Philadelphia Story* in Manhattan's East Village, lost in thought, I walked past a panhandler lurking on the corner, who—completely unprovoked, I might add—snarled after me, "Ice queen!" The epithet stung. It bothered me for years, precisely because it was such a shock to realize how differently a complete stranger had perceived me. His impression of me was nothing like the image I had internalized as my true self. I was a painfully shy, struggling young writer who had just been swept away by the crackling on-screen chemistry between Katharine Hepburn (Hollywood's ice queen) and Cary Grant. The panhandler only saw a stuck-up, frigid white girl who couldn't spare a dime. I realized this was a common perception among people who didn't know me well—that I was cold, aloof, unfriendly—and it was one I felt helpless to change.

So it was reassuring to hear Vazire admit that she, too, struggled with a troubling mismatch between her self-perception and how others perceived her. Vazire is young, attractive, highly intelligent, and has attained an impressive level of achievement very quickly in her chosen field. She is also shy and reserved, speaking in quiet, measured tones, and she rarely smiles. Her demeanor initially caused a certain amount of collegial friction until she learned how to compensate for it. "Shy people, especially attractive shy people, are almost always seen as arrogant and cold on a first impression," explained Erika Carlson, one of Vazire's graduate students. "People get angry at you for not talking." They take your silence personally, as a reflection on them. Vazire's early academic success exacerbated the effect.

So what can you do? "When people have a big mismatch between the inner self—the self they believe themselves to be—and how other people view them, it can be enormously stressful," said another WUSTL colleague, Michael Strube. So we will seek to change that other person's view and bring it in line with our own. "We are constantly trying to create consistency between the views we have of ourselves and the views that others have of us, because it just makes for more harmonious social interactions." The solution is not rampant homicide, as Debbie Jelinsky would have you believe, but transparency: you need to help others see the self you have internalized.

Now when Vazire meets new people, she makes an effort to assure them that if she is not speaking much, it is not because she is cold or disinterested, but merely because she is quiet. Over time, I became much better at expressing my inner warm, affectionate nature. It requires a willingness to be open and vulnerable, which is difficult for shy people, given our deep-seated fear of social ridicule and embarrassment. But the benefits far outweigh the risks of potential humiliation. Perhaps we can't change our basic temperament, but we can use the feedback we receive from others to finesse our self-presentation and smooth our social interactions.

This is just one of many ways in which unique environment shapes the character aspect of our personalities, but are psychologists correct that our basic temperaments are determined by our genes? Identical twins share all DNA and, when raised together, most of the family environment; fraternal twins share half of their DNA and most of the family environment, which means by analyzing the differences between the two sets, one can determine how much of a given trait is heritable. Identical twins are much more similar to each other in terms of personality traits than fraternal twins are, and this is true even if those twins were raised apart. Non-twins who are adopted, like me, tend to resemble their biological siblings when it comes to personality traits, not their adoptive siblings, despite having no shared family environment with the former.

This strongly suggests that the Big Five personality traits are at least partly heritable, and scientists have been able to estimate just how much. Those estimates range from 42 percent for agreeableness to 57 percent for openness. As for the rest, one's unique environment—as opposed to one's shared familial environment—turns out to be far more influential. "The determinant of where we fall [on the Big Five scale] is how our brains are wired up," Daniel Nettle writes in *Personality*. "And the determinants of how the relevant parts of our brains get wired up are firstly genetics, and secondly, various early life influences over which we have no control and which seem essentially irreversible."

The Long and Short of It

"I write of melancholy, by being busy to avoid melancholy," a humble Oxford vicar observed in 1621. "There is no greater cause of melancholy than idleness, no better cure than business." The vicar was Robert Burton, author of the classic *Anatomy of Melancholy*, which boasts one of the longest subtitles in literary history: *What It Is: With all the Kinds, Causes, Symptomes,*

Prognostickes, and Several Cures of It, in Three Maine Partitions with their Several Sections, Members and Subsections. Philosophically, Medicinally, Historically, Opened and Cut Up. In his examination of one of the most common psychological maladies of the era, Burton uses the medieval notion of "melancholy" (the condition we now call clinical depression) to explore a broad range of scientific and philosophical topics, with entertaining digressions into goblins, digestion, and American geography.

Despite possessing a wicked sense of humor—Burton purportedly found endless amusement in the cursing and quarreling of the local bargemen—he also suffered from chronic depression. In fact, he was rumored to have hanged himself in his Christchurch chambers in 1640. He viewed his condition as "a habit, a serious ailment, a settled humour . . . not errant but fixed," opining that therefore "it will hardly be removed." So Burton was keenly interested in the subject of melancholy, which he naturally attributed to an excess of black bile. Modern science, however, has pinpointed a different culprit: a neurochemical called serotonin that appears to be linked to the trait of neuroticism.

Serotonin is a simple molecule that comes from tryptophan, an amino acid found in foods that is often blamed for that sluggish, relaxed feeling after heavy meals. It is regulated by a complex system that affects several major regions of the brain, with axons snaking from the limbic system into the cerebral cortex and frontal lobes, and offshoots into the hippocampus and pituitary gland, as well as the amygdala (the "fear center" of the brain). Taken together, they act as the brain's "alarm system," responding to outside stimuli, and some people's alarms are set at a higher sensitivity than others. Brain scans have shown that people who score higher on neuroticism are more responsive to negative stimuli. There are also differences in the size or density of the amygdala in those who score higher on neuroticism, as well as lower activity and density in the hippocampus and right frontal lobe.

When sensory stimuli trigger the neurons in these regions, serotonin floods into the spaces between the neuron cells. There, it either starts a chemical reaction, such as releasing the anxiety-causing hormone cortisol into the brain, or it gets taken back into the cell, a process called reuptake. The protein responsible for that reuptake is called a transporter. Some people produce serotonin transporters that are more efficient at funneling serotonin back into the cells. Because serotonin affects so many different regions of the brain associated with emotional response, sensory perception, impulse control, and empathy, your serotonin receptors can affect all kinds of behaviors and personality traits. That's why the antidepressant Prozac targets serotonin transporters.

Genes provide the blueprint for the serotonin system, and that system is largely in place when we are born. There are a dozen different kinds of serotonin receptors—the molecules that line cells and bond with only specific neurochemicals— each with its own distinct gene, but there is only one known serotonin transporter determined by just one gene, what geneticist Dean Hamer likes to call the "Prozac gene." There is one particular section of DNA in this gene that, when removed, increases the activation of the gene. So the function may involve slowing down that activation process. This gene has an unusual sequence of sixteen repetitions. We all have a gene that makes serotonin transporter, but some people have a version with the full sixteen copies, while others have a shorter version with only fourteen copies.

The longer version makes a lot more of the serotonin transporter than the shorter version. Roughly one-third of the human population has two copies of the long form of the gene, while two-thirds have either one or two copies of the short version. The short version is dominant, so those with that variant make less of the serotonin transporter proteins, which means they will be more anxious than their long-form counterparts. Those with the highest anxiety levels have the short version of the gene. There is one caveat: neuroticism is roughly 48 percent heritable, yet those variations in the serotonin transporter gene account for only 3 to 4 percent of the total observed variability in the general population. There could be more than a dozen genes involved in this particular aspect of temperament; we just haven't found them yet.

Serotonin and Hamer's "Prozac gene" also plays a role in another trait associated with neuroticism: harm avoidance, which can also be linked to being shy or fearful, as well as anxious and depressed. Brain scans showed more activity in the right frontal area in shyer children, and more activity in the left frontal area for more outgoing children. Since the right side deals with controlling negative emotions, while the left is more involved with positive emotions, it makes sense that the children's temperament reflected which side of the brain was dominant. Sampling the saliva of shyer children also revealed twice as high levels of the stress hormone cortisol.

As a child I was painfully shy and timid, only gradually learning to relax and open up. I was born that way—that is, with a genetic predisposition to higher levels of anxiety and a strong propensity for harm avoidance, the essence of shyness. That meant that negative events during my childhood affected me more deeply than a child with the long form of the gene, who would be more likely to bounce right back and gain confidence from the same experiences. So that initial tendency was reinforced. Yet that doesn't mean I was doomed to go through life with crippling anxiety. In *Living with Our Genes*, Hamer tells the story of a shy woman who found herself in a career that forced her to become more socially engaged. Over time, as she pushed herself to overcome her natural reticence and shyness, she gradually lost her fear of such situations. The same thing happened to me. We can choose to pursue experiences that reshape the character aspects of our personalities, even if our basic temperament remains unchanged.

Many people confuse shyness with introversion, but the degree to which one is introverted or extroverted is linked to a different neurochemical: dopamine, which is derived from an amino acid called tyrosine and is tied to the brain's reward system, most notably the nucleus accumbens: the pleasure center of the brain, or the brain's "G-spot," as Hamer likes to call it. Along with the prefrontal cortex, sensory cortex, and motor region—all of which appear to be linked to extroversion, especially the traits of impulsivity and novelty-seeking behavior—the nucleus accumbens has neurons with a large number of dopamine receptors.

Receptors are like locks: large proteins sitting on the surface of brain cells that respond to only certain neurochemicals (the keys). Whatever transmitter acts as the key—in this case dopamine—binds to the receptor molecule and initiates a flood of chemical reactions. This is the "rush" or "buzz" we get from pleasurable activities, and it can be measured with fMRI. There is a markedly pronounced increase in activity in the relevant brain regions in highly extroverted people, notably the limbic regions associated with emotional response to stimuli. The response was much smaller in more introverted subjects. According to Daniel Nettle, because extroverts get a bigger rush from social contact, thrill-seeking, or other activities, they are far more motivated than introverts to put in the effort to seek out such things, thereby reinforcing extroverted behavior.

Like the serotonin transporter gene, one of the genes that codes for dopamine receptor molecules comes in both long and short variants. These are pegged to a repeating sequence of forty-eight pairs of DNA bases. The most common variants have four and seven variants, although other versions run the gamut in between. The long version has more than six repeats and appears to be dominant; the short version has less than six repeats. People who have at least one copy of the long version tend to score higher on extroversion and novelty-seeking. The longer the gene, the greater the need for novelty. However, there may be hundreds of genes involved in dopamine-signaling pathways. Novelty-seeking is roughly 40 percent heritable, with this particular gene accounting for about 10 percent.

When it comes to agreeableness (or the lack thereof), we can pinpoint two hormones in particular, oxytocin and vasopressin. In 2011, UCLA scientists found an oxytocin receptor gene that seems to be linked to a lack of optimism, low self-esteem, and little belief that one has control over one's own life (agency). There are two versions of the gene: one with an A (adenine) variant and the other with a G (guanine) variant. People who have at least one A variant may be more vulnerable to stress and have poorer social skills. The UCLA study found that people with one or two A variants at this specific location on the oxytocin receptor gene showed significantly lower levels of optimism and self-esteem, and were far more likely to be depressed, than people with two G nucleotides. Naturally, the researchers caution that this is just one factor among many that can influence one's psychological outlook; others include a supportive childhood, good relationships, close friends, and other genes.

Oxytocin is often called the "cuddle chemical" or the "love drug," both rather nauseating nicknames stemming from its role in fostering social bonds between people, making them more caring, affectionate, or generous—but only with members of their own tribe. It makes them react harshly to perceived outsiders, which is why psychologist and author Carol Tavris has described oxytocin as the "cuddle your own kind and to hell with the rest of you" chemical. In 2012, psychologists at the University of Buffalo and the University California–Irvine announced they had found a "niceness gene"—not a gene that makes you nicer, but a gene that codes for key receptors for oxytocin and vasopressin.

Certain variants of that gene produce receptors that are better at binding those hormones to cells. It is not the only gene that influences these two hormones, but intriguingly, this variant only predicted the likelihood of human behavior in combination with people's experiences, which shapes their perceptions about the world. People who were more inclined to view the world as a threatening place, and who also had this "niceness gene," were more likely to overcome their fears and exhibit generosity in helping others. Those with the same worldview who didn't have the gene were far less generous. The "booster gene" didn't have any effect on those already inclined to view the world positively.

Serotonin, dopamine, oxytocin, and vasopressin are just the primary culprits known so far when it comes to genetic influence on personality; there are likely to be countless other factors yet to be discovered. The point is that we are not blank slates, but neither are we slaves to our genetic destiny. Our genes influence our personality by regulating brain chemistry, and that in turn influences how we perceive the world and react to that sensory input. There is still plenty of built-in flexibility to enable us to adapt and evolve in response to changing circumstances. "Genes are not fixed instructions," Dean Hamer writes, likening them not so much to a musical score, but to musical instruments. "Genes don't determine exactly what music is played—or how well—but they do determine the range of what is possible." That is far more influence on our personalities than we get from humors, blood type, or the stars.

II

MYSELF

4

Three and I'm Under the Table

A man who drinks too much on occasion is still the
same man as he was sober. An alcoholic, a real alcoholic, is not the same man at all. You can't predict
anything about him for sure, except that he will be
someone you never met before.

—RAYMOND CHANDLER, *The Long Goodbye*

The fruit flies under the microscope were so motionless that
for a moment I suspected they were dead. Then I noticed
the slight twitching of antennae and the occasional kick from a
tiny foreleg. The flies were merely in a deep, deep sleep—an
alcohol-induced slumber, brought on by a few too many happy
hours in Ulrike Heberlein's "Inebriometer," a four-foot vertical
glass tube full of ethanol vapor. It is their own private cocktail
lounge, where they can indulge to their heart's content.

And indulge they do. "They drink a lot," Heberlein admitted, although they don't have much choice when it comes to
breathing in the vapor. She estimates that her fruit flies can im-
bibe half their body volume in 15 percent (30 proof) alcohol
each day. Not even spiking the ethanol with bitter quinine slows

them down. Granted, they have ultrafast metabolisms—the better to burn off the booze—but they can still become falling-down drunk, toppling down the tubes of the Inebriometer as they lose coordination and stumbling around the bottom before passing out. Heberlein determines how drunk they are by timing how long it takes for them to lose consciousness. Like humans, fruit flies vary in their tolerance for alcohol—that is, how quickly their bodies adapt to the ethanol, requiring them to inhale more and more of it to achieve the same physical effects. The secret behind this might just be hiding in their genes.

Heberlein is a behavioral geneticist at the University of California–San Francisco, where she applies genetic and molecular analysis to the fruit fly *Drosophila melanogaster*. Essentially, she gets fruit flies drunk for a living. She knocks out various genes that regulate their tolerance for alcohol and studies how this changes their behavior under the influence. The batches of flies she breeds for this purpose all have colorful nicknames like "Barfly," "Hangover," "Tipsy," "Lightweight," "Happy Hour," and "Cheap Date." What else would you expect from a scientist whose office window boasts a plastic fruit fly holding a martini glass?

The Chilean-born Heberlein confessed that when she named that last class of flies, she didn't fully comprehend the sexual implications of the slang: "I just thought 'cheap date' meant someone with a low tolerance for alcohol." But the slightly salacious moniker garnered attention for her research, and the name has stuck.

I admitted to having a low tolerance for alcohol, generally limiting myself to one or two drinks with dinner, ever since an unfortunate college experiment mixing many different liquors—because how would I know which drinks I liked if I

didn't sample everything? That ill-advised research left me retch-
ing until the wee hours, before curling up in the fetal position
on the cool bathroom tile, waiting for the room to stop spinning
so I could sleep. (I believe the polite euphemism is "conversing
with the porcelain Buddha.") It took me two days to recover
fully, and I never binged like that again.

Heberlein laughed at my story. "Oh, you are Cheap Date!"
she exclaimed. Yes. Just like her fruit flies. Or rather, since there
is no known correlating gene to Cheap Date in humans yet, I
very likely have a gene (or genes) with a comparable effect on
my tolerance for alcohol.* While our culture tends to lionize
those who can hold their liquor and drink lesser mortals under
the table, my low tolerance does confer one advantage: it re-
duces my odds of becoming an alcoholic. I simply can't pound
back the cocktails fast enough, night after night, year after year,
to develop a physical dependence on the substance.

The National Institute on Alcohol Abuse and Alcoholism
(NIAAA) estimates that 17.6 million Americans—one in every
twelve adults—either abuse alcohol or are alcohol dependent.
According to the clinical definition, telltale signs of alcoholism
include a powerful urge to drink (cravings); the inability to stop
after just one or two drinks (loss of control); withdrawal symptoms such as nausea, sweating, and the shakes (an indication of
physical dependence, commonly known as delirium tremens, or
the DTs); and the need to drink ever greater amounts of alcohol
to reach the desired level of inebriation (high tolerance).

Based on those criteria, I am the anti-alcoholic. While I can

---

* There is a mammalian equivalent of Cheap Date that has been found in
mice, and mice with mutations in that gene exhibit altered behavior when
consuming ethanol.

savor a well-made sidecar or a nice vintage Châteauneuf-du-Pape, I don't crave alcohol. I often go for long periods without imbibing, don't suffer withdrawal when I do so, and have no problem stopping after one or two drinks—an ability that has astonished and frustrated recovering alcoholics I know by depriving them of the chance to drink vicariously through me. But I can't take too much credit for the breezy ease with which I manage this austerity; it's as likely due to the luck of the genetic draw as it is to my own self-discipline.

Alcoholism has a strong hereditary factor. The children of alcoholics are nearly four times more likely to develop drinking problems. Multiple studies of identical twins have shown that if one identical twin is an alcoholic, there is a 76 percent chance the other is too. For fraternal twins, that probability drops to 26 percent, still well above the rate of alcoholism in the general population. This holds true even for adoptees. The Stockholm Adoption Study, for example, found that if an adopted man's biological father was an alcoholic, that adoptee was six times more likely to become one as well, even if raised by non-alcoholic parents.

With the successful sequencing of the human genome, scientists have begun to home in on specific genes that might affect whether a given individual develops a drinking problem. That said, we still can't point to one specific gene as the primary cause for alcoholism. Behaviors cannot be reduced to traits. "All behaviors are very complex," Heberlein explained. "There's no such thing as an alcoholism gene. There are genetic risk factors." There are dozens of genes that influence one's likelihood of becoming an alcoholic, and none act alone, which is why Heberlein rejects the notion of purely genetic determinism. Alcohol dependence is between 38 percent and 64 percent heritable, with a constant give and take between genes and environmental influences. A family history of drug or alcohol abuse, poverty, peer pressure, cultural attitudes toward alcohol, stress, anxiety, depression, or certain personality traits can all contribute to tip the balance one way or the other. So when it comes down to the central question—are alcoholics born or made?—science equivocates by answering truthfully: "Eh, it's a bit of both, actually."

Tiny Tipplers

Watching fruit flies get drunk can be surprisingly entertaining. Their reactions largely mirror our own. Initially, they become hyperactive, as people often do after one or two drinks to "loosen up." Once the alcohol concentrations in their tiny bodies start to increase, the flies fall over and bump into one another, resembling humans with slurred speech and diminished motor coordination. Male fruit flies also become a bit randy, although they are less likely to mate successfully—perhaps because, in their alcohol-induced haze, they frequently court other males by mistake, instead of females.

Eventually the flies become so sluggish, they tip over and fall asleep—much like that buddy who has a bit too much to drink at a party and suddenly finds the pavement outside so very comfortable that he decides to take a nap right there. Fruit flies even experience their own version of the DTs, indicating they can, indeed, become micro-sized alcoholics.

"Once I saw a drunk fly, I just became so fascinated by the behavior," Heberlein said as she showed me around her state-of-the-art laboratory in a sleek, modern building on UCSF's Mission Bay campus. Fruit flies are the workhorses of modern genetics research, used to study everything from cancer to sleep disorders; they make excellent model systems because we share nearly two-thirds of our genes with the critters. They are cheap, easy to breed, and their genes can be manipulated easily. They are also exceedingly fond of fermented foodstuffs, although technically it is the yeast they seem to crave—if only as a navigational guide to delicious rotting fruit. Heberlein noted that if someone in her lab opens a bottle of beer, the fruit flies swarm toward the alcohol, jumping in with wild abandon—and more often than not, drowning in the process.

Heberlein is trying to pinpoint the specific genes that regulate boozy behavior. She compared the process to figuring out how a car engine works by removing one part at time and then evaluating how its removal affects the functionality of the engine as a whole. Ideally, one can put the genetic "engine" back together with a clearer understanding of how all those parts combine to make up the whole.

There are thirty or forty different genes that play some kind of role in how the flies respond to alcohol. The Cheap Date flies have a gene that regulates sensitivity to alcohol. Flies with the Tipsy gene pass out after lower doses of ethanol, while those with the Barfly gene can really hold their liquor, as can flies with mutated Happy Hour genes. The Barfly gene, when intact, inhibits the effects of a protein called epidermal growth factor (EGF) that is linked to reducing tolerance for alcohol. The mutated version in Heberlein's Happy Hour flies has the opposite effect; the influence of EGF is greatly diminished, resulting in a higher tolerance.

It is not sufficient to simply catalog the various genes and determine how they work, Heberlein insisted. "We also need to know how they interact before we can really understand what is happening." We think of a gene as something that encodes a trait, but a single gene might produce many different proteins, each serving different functions. Furthermore, experiences change the expression of the gene by modifying the proteins that package the DNA, or by modifying the DNA directly. Stress, for instance, can leave a molecular signature in the genome, thereby changing an organism's behavior.

The Hangover gene is also linked to higher tolerance in fruit flies. When Heberlein placed normal flies in the Inebriometer and let the flies get good and drunk, they took about twenty minutes to pass out. Flies recover in about fifteen minutes or so. After four hours of sobriety, Heberlein put them back into the Inebriometer for another round. This time, those flies with a functioning Hangover gene took longer to pass out— twenty-eight minutes—indicating an increased tolerance for alcohol, and they also had to take in more alcohol to achieve the same level of intoxication. Flies whose Hangover gene was inactive still took only twenty minutes to pass out during their first Inebriometer ride. But when tested four hours later, they passed out after only twenty-two rather than twenty-eight minutes. They didn't build up a tolerance as high as that of the normal flies.

Heberlein then literally turned up the heat with a new batch of flies, increasing the temperature in their habitat until it triggered a stress response. When those with functional Hangover genes were placed in the Inebriometer four hours after thirty minutes of sweating it out the heat, they showed a greater tolerance than normal, taking twenty-nine minutes (instead of twenty) to pass out, even though it was their first exposure to alcohol. This didn't happen with a batch of flies with a non-working version of the Hangover gene. For Heberlein, that suggested the same gene controls the flies' response to stress, consistent with studies in humans demonstrating a similar link between stress and alcohol tolerance.

Romantic rejection can also drive *Drosophila* to drink. Fruit flies have a remarkably elaborate courtship ritual. The male fruit fly usually spots the object of his desire near the feeding area and follows her at a respectful distance until she indicates receptiveness, as a human male might send over a drink to break the ice. He then gently taps her with his foreleg. If she responds with the appropriate chemical signal, he launches into the fruit-fly courtship song, vibrating his wings in what one can only assume is the *Drosophila* version of Barry White. If she is sufficiently impressed with the performance to let him lick her abdomen, the deal is all but sealed, and they go on to mate. All that's missing is the mood lighting. "I mean, talk about fore-play!" Heberlein marveled.

But there is a catch. Once a female fruit fly has mated, she will (at least temporarily) reject vehemently any further advances from male fruit flies. Unlike the playful stages of fore-play, "It's pretty nasty to watch," Heberlein admitted. "She kicks him, she runs away, she shoves her ovipositor in his face." There is only so much abuse a male fruit fly can take before he loses interest in courtship rituals entirely, even with other virgin females. His decreased libido can last up to a week.

Such mating behavior illustrates why fruit flies are so useful for studying learning and memory, but Heberlein was interested in exploring what she terms "social defeat." She paired virgin males with mated females for an hour at a time, three times a day, for four straight days, subjecting them to constant romantic rejection. Then the long-suffering males were placed in an alcohol-drinking assay, where they proceeded to drown their sorrows by drinking more than twice as much alcohol as the lucky lotharios in the control group that had successfully mated. Once the rejected males were allowed to mate, they went back to drinking less alcohol when they returned to their little bar.

It is tempting to anthropomorphize the fruit flies and draw parallels to similar reactions to the experience of romantic rejection among human males (or females). When Heberlein described her work to a small group of Hollywood insiders, the first question from a mischievous producer was, "Please, tell us more about how sex with virgins can cure alcoholism."* But this is projection; fruit flies are simple organisms, with tiny brains, incapable of the kind of complex emotions routinely experienced by humans.

The first question a *scientist* asks in such a case is, "What's the mechanism?" For the fruit flies, Heberlein suspected that some part of the act of mating triggered the reward response in their tiny fruit-fly brains. But how to demonstrate that fruit flies found the act rewarding? You can't ask them directly. Instead, Heberlein's team trained both mated and virgin male fruit flies to associate two different odors with mating and lack of mating, respectively, and then waited to see which of the two odors they preferred.

When a male fruit fly experiences rejection and fails to mate, it decreases the levels of a small molecule called a neuropeptide, involved in normal communication in the brain. Levels increase when flies mate successfully and decrease after rejection. Heberlein bred flies with lower levels of the neuropeptide, allowing them to mate before placing them in the drinking assay. They

---

*The event was a salon hosted by director Jerry Zucker and his wife, producer Janet Zucker, in partnership with the Science & Entertainment Exchange, an outreach program of the National Academy of Sciences.

proceeded to drink at the much higher levels more typical of rejected males. The opposite also proved to be true. When He-berlein activated the same circuit in rejected males, she found they drank reduced amounts more typical of mated males, as if they had not been rejected.

As fascinating as drunken fruit flies might be, it is tough to draw direct correlations between the genes of a fly and those of humans. That is why Heberlein started scaling up her studies to mice. Mice are mammals and have 80 percent of their genes in common with humans, so they are a logical next step in terms of animal models for alcoholism. She employs the same ap-proach as with her fruit flies: she knocks out one specific target gene in a single generation and then studies how this changes their drinking behavior.* As with mice, so with men: just like Heberlein's fruit flies, some humans appear to be genetically predisposed to have high (or low) tolerance for alcohol.

Of Mice and Men

One summer afternoon in 1932, Tony Marino, the owner of Marino's speakeasy in the Bronx, along with a few cohorts, came up with a brilliant scheme to save his failing business. They would take out life insurance on one Michael Malloy—a hard-drinking Irishman in his sixties—and let him drink him-self to death, splitting the payoff between them. The plan seemed foolproof. Malloy had no friends, no known family, no steady job, and showed up at Marino's every morning, drinking

---

* So far, Heberlein has found that eight of the ten genes tested using fruit flies show the same effects in mice. Collaborating with other geneti-cists revealed similar changes in human genes associated with response to alcohol.

steadily until he passed out on the floor—or until Marino cut him off. Nobody would miss him, and he was arguably half dead from drink anyway.

So the members of the "Murder Trust," as they came to be known, set the scheme in motion. Once the policies had been secured, Marino offered Malloy an open tab and started pouring straight shots of whiskey and gin. Malloy downed them all without batting an eye, wiped his mouth, and promised to be back for more. This went on for three days, with no noticeable ill effects, and the Trust became impatient. So they started serving Malloy straight wood alcohol—stuff so toxic even just a little could cause blindness. It was the early days of the Prohibition era. Wood-alcohol poisoning had killed more than 50,000 people in the United States by 1929. Malloy downed that, too, night after night, and still came back for more.

By now the Trust was getting desperate. They tried serving him a sardine sandwich mixed with metal shrapnel. Nothing. They stripped off his shirt, doused him with cold water, and left him on a park bench on a cold winter night, hoping he would freeze to death. He was back at the bar the next day, complaining of a "wee chill." They hired a cabdriver to run Malloy down in the street, backing over his prone body for good measure. He showed up five days later after a brief hospital stay, battered and bruised, and cheerfully announced, "I sure am dying for a drink!"

It took forced asphyxiation via a rubber tube running from a gas-lamp fixture to his mouth to finally finish him off. Even then, Malloy had the last laugh. A suspicious insurance agent didn't buy the forged death certificate listing lobar pneumonia under cause of death, and launched an investigation. In the end, the founding members of the Murder Trust were convicted of first-degree murder and summarily executed.

That, my friends, is tolerance for alcohol on a grand scale. Malloy arguably hit the genetic jackpot when it came to holding his liquor, bolstered by decades of hard-core drinking. At Indiana University and Purdue, geneticist Nicholas Grahame thinks his collection of binge-drinking mice could give Malloy a run for his money.

Grahame studies genetics through selective breeding of mice, a method that's been around practically since humans formed agricultural societies. Unlike Heberlein's transgenic mice, in which one gene changes in a single generation, successive generations of Grahame's mice have been bred over the last fourteen years to show a slight preference for alcohol. In essence, he and Heberlein are studying the genetic underpinnings of the same behaviors from complementary micro and macro perspectives.

"Alcoholism is not defined by a genetic trait; it is defined by a behavior: drinking to excess," Grahame said about his preference. "So to me, the best way to look at the genetics of alcoholism is to be blind about what genes you are changing and to focus on the behavior instead." Natural selection changes many genes, not just one at a time. This approach gives him the advantage of larger and more easily detectable behavioral changes. But it also carries a disadvantage: uncertainty. "I don't always know what I've done at the genetic level," Grahame explained. So he also breeds a control group of teetotaling mice that hate alcohol as much as the boozy group loves it. This lets him compare the genes, brains, and behaviors of the two groups to determine any genetic signatures associated with excessive drinking.

It took forty generations of incremental genetic changes to produce Grahame's tippling rodents, because unlike fruit flies and the hardy Michael Malloy, mice do not have a natural affinity for the stuff. They just don't like the taste, choosing water over alcohol almost every time in laboratory tests. Oddly enough, though, "mice really seem to enjoy being drunk," said Grahame. If you inject a mouse with sufficient alcohol to get the mouse drunk in a particular location, it will later show a marked preference for that location over other sites where it was just hanging out, bored and stone-cold sober.

It is no mean feat to get mice drunk, even this new booze-guzzling breed, because their speedy metabolisms churn through the stuff at eight times the rate of a human being. Grahame gets around this by spiking their water with 10 percent alcohol, which means they are boozing it up at roughly 1.5 to 2 kg per hour. Since they metabolize alcohol at 1 kg per hour, they eventually become intoxicated. They will keep drinking, too, hitting blood alcohol levels three times higher than the equivalent of the legal limit for humans. It's on a par with a 190-pound man knocking back more than three gallons of wine per day. Not surprisingly, the mice performed badly when he placed them on mini balance beams to assess their motor control. But the intake didn't slake their thirst for booze one whit. "They got drunker than any other animal model that has ever been studied," Grahame said with a hint of pride. He has genetically engineered the ultimate binge-drinking machines.

One could argue that nature did this whole selective-breeding experiment already with Michael Malloy. He is long dead, but there is no shortage of chronic heavy drinkers—and their long-suffering relatives—available for human studies. Animal models help narrow the field of genetic candidates and provide scientists with valuable clues as to where to look on our own genome—scientists like Danielle Dick, a psychologist at Virginia Commonwealth University.

In 2004, Dick pegged one specific allele in humans that appears to be linked to alcoholism. It is one of several receptors for gamma-aminobutyric acid (GABA), a chemical neurotransmitter that regulates stress and anxiety in the central nervous system. Sample groups in such studies include both alcoholics and non-alcoholics. The researcher scans the genomes of all participants, looking for points where there is a shared genetic variation among the alcoholics that isn't present in the non-alcoholics. Dick and her colleagues have studied the DNA of more than 10,000 donors from families with particularly strong patterns of alcoholism and found small genetic differences in those more prone to drinking problems.

It is still unclear what role this allele plays in one's predisposition to drink, but Dick hypothesizes that it might be related to brain overactivation. GABA works by letting more chloride ion into brain cells to inhibit the firing of neurons. Drugs like diazepam increase GABA's effectiveness, while caffeine (a stimulant) does the opposite. Alcohol, like diazepam, has a sedative effect, and regular heavy drinking may increase the number of GABA receptors in the brain, increasing one's tolerance even as it soothes anxiety. It's possible that people with a hyper-excited central nervous system are, in essence, self-medicating. That puts them at greater risk of developing alcohol dependence.

What does my own genotype have to say about my propensity toward alcoholism? Overall, I have a "slightly elevated risk." That assessment is based on the handful of studies 23andMe can cite with good confidence showing measurable correlations. By now we have all memorized the mantra, "Correlation is not causation," yet correlations can provide vital clues about where to look. One of those studies pertains to the prevalence of alcohol dependence in men. Ergo, although I carry this particular genetic variant for "slightly higher odds" than the mean population, I am female, so this is unlikely to be a relevant factor.

Ditto for another genetic variant, common in those of Asian or Ashkenazi Jewish descent, which causes a severe flushing response to alcohol. This is due to a deficiency in an enzyme our bodies need to process acetaldehyde, a toxic carcinogen produced while alcohol is broken down and metabolized. The enzyme turns acetaldehyde into the relatively harmless acetic acid (vinegar), so a deficiency results in a nasty buildup of the toxin in the body, leading to the aforementioned flushing effect. Those with the bad luck to inherit two copies of the variant (one from each parent) are particularly sensitive, experiencing nausea and a racing heartbeat for good measure. Since I am not of Asian or Ashkenazi Jewish descent, it is unsurprising to find I do not have this gene.

My slightly elevated risk factor comes from a dopamine receptor allele. Dopamine is the so-called pleasure neurochemical, a reward mechanism in the brain that is key to the fun humans find in eating, drinking, and sex. It may also be linked to addiction. In 2007, researchers analyzed the data from 10,000 participants collected in earlier studies and found an increased risk of alcoholism for every copy of this particular allele (A) that is inherited. If both parents provide a copy of the allele (AA), one's risks of alcoholism are significantly higher; if both parents provide a different version of the allele (GG), the risk is typical of the mean population. I fall in between (AG), receiving just one copy of the offending allele—hence the slightly elevated risk.

Should I ever develop a dependency and find myself in detox, it probably won't be pleasant, since I also did not inherit either of the protective genetic variants that would reduce the likelihood of my suffering seizures during withdrawal. Fortunately, I

seem to have plenty of other genes that provide me with a pa-
thetically low tolerance for alcohol, limiting my alcohol consumption and thereby offsetting the minor genetic risk factors I
inherited from one of my biological parents. And remember that
genes alone do not fully account for whether you become an
alcoholic; the expression of those genes is influenced by many
other factors.

True Confessions

Members of London's literary intelligentsia were electrified in
1821, when *London Magazine* published a two-part anonymous account of one man's addiction to laudanum—a heady
mix of opium and alcohol. *Confessions of an English Opium
Eater* spared no detail in its account of the pleasures and perils
of the drug. William Burroughs, who knew something about
the subject, enthused, "No other author since has given such a
completely analytical description of what it is like to be a junkie
from the first use to the effects of withdrawal." That author was
the essayist Thomas De Quincey, who spent his life popping
pills out of a handy snuffbox, dodging creditors, and occasionally producing short pieces of literary brilliance that filled fourteen volumes after his death in 1859.

De Quincey seemed to think his upbringing was at least partially to blame for his addiction; he cataloged in loving detail
the emotional and psychological traumas he claimed to have
suffered in his youth, most notably his horrific experiences as a
homeless teen runaway. But on the surface, there was nothing
remarkable about his early childhood. He was born in Man-
chester, England, to a successful merchant, although his father
died when he was nine, at which point the family moved to

Bath. Still, they were hardly destitute, and his mother, though strict, made sure her family lacked for nothing.

It is true that young Thomas was a sickly child: there are references to digestive problems, astigmatism, and severe facial pain (neuralgia), and at least one scholar has speculated that he may have had a mild case of infantile paralysis at some point in his life. Timid and prone to bouts of depression, De Quincey spent a lot of time alone as a child, but he was also highly intelligent, and an avid reader. "That boy could harangue an Athenian mob better than you or I could address an English one," his headmaster once declared.

The plan was for De Quincey to spend three years at the equivalent of a prep school and earn a scholarship to Oxford University, but the young man bolted after just nineteen months, wandering through Wales on his own before returning to London—anything to avoid his family, despite being on the brink of starvation. Eventually he enrolled at Worcester College and completed his coursework, but he never graduated. By then he was already taking opium, ostensibly to relieve his facial neuralgia, and within a year he was using the drug daily for recreation. Despite the phenomenal success of his *Confessions*, addiction and debt would plague him for the rest of his life.

Was De Quincey correct in blaming his unhappy childhood for his adult addiction? Not necessarily. George Vaillant conducted two landmark studies that followed six hundred American men over several decades, tracking their drinking habits from youth to old age, and gathered the results in his 1995 book, *The Natural History of Alcoholism Revisited*. He found that unhappy childhoods were not a significant factor in whether a man became an alcoholic—unless, of course, that bad home environment existed because someone else in the family was an alcoholic, evidence of a hereditary rather than environmental influence. But there were certain personality traits that seemed to correlate with heavy drinking later in life. While the classic alcoholic personality traits—self-centeredness, immaturity, resentfulness, irresponsibility—emerged after a man had started drinking, Vaillant found that those men with extroverted or antisocial (sociopathic) personalities were most likely to become alcoholics.

One could speculate endlessly about De Quincey's psychological problems, but today he may well have been diagnosed with antisocial personality disorder (APD). De Quincey had some early risk factors—the loss of a parent and chronic depression—and he certainly felt he'd been subjected to verbal or physical abuse of some kind as a child. He was charming, manipulative, frequently agitated, and careless about deadlines. He quarreled regularly with his long-suffering editors and had repeated run-ins with the law, including a stint in debtor's prison. His *Confessions* was notable for its lack of remorse; critics felt the book glamorized addiction despite the vivid accounts of the physical and mental horrors of withdrawal. Above all, De Quincey was reckless and impulsive, characteristics associated with both APD and drug and alcohol addiction, according to Nicholas Grahame, who estimates that about half of those diagnosed with APD are also alcoholics.*

Impulsive behavior includes not considering the consequences of actions in favor of immediate gratification, as well as sensation-seeking. There is even a gene linked to such behavior. Grahame has modeled impulsivity in his mice by giving them

—————————————

*Grahame is quick to point out that the two are not synonymous, merely that they share some of the same personality quirks.

the rodent version of a "delayed discounting" test. In humans, the test involves a series of questions: Would you rather have $20 now or wait one year and get $100? What if it were $50 right now and $100 in a year? What about $70 right now and $100 in a year? "At some point you switch your choice from the delayed reward to the immediate reward," Grahame explained. People who are alcoholics show a marked preference for immediacy. In the case of mice, he offered a sweet drink now, or more if the mice waited eight seconds. ("Mice are not very patient.") True to form, his binge-drinking mice showed a stronger preference for immediate reward compared to mice that were bred to not drink alcohol.

Impulsivity shows up surprisingly early in life. In 1972, Walter Mischel conducted a landmark behavioral study with more than six hundred children between the ages of four and six at Stanford University's Bing Nursery School. Each child was given a marshmallow and told they could eat it immediately if they chose. But if they opted to wait for fifteen minutes, they would be given a second marshmallow as a reward. Mischel would leave the room and a hidden camera would record the children's behavior.

The results were sometimes comical, and often poignant, as each child struggled to resist temptation. Some children just ate the marshmallow right away, while others tried to distract themselves by covering their eyes, kicking the desk, or poking at the marshmallow with their fingers. Some picked up the marshmallow and smelled it, or licked it, before giving into temptation and cramming it into their mouths. Others took little nibbles around the edges. In the end, approximately one-third of the children managed to hold out long enough to win the second marshmallow.

But the real significance of Mischel's experiment was what he discovered several years later. His own daughter had been a classmate of many of the original test subjects, and through her he heard how some of them turned out. There seemed to be an unexpected correlation between the success of some of those children later in life—better grades, greater self-confidence, healthier relationships, more advanced academic degrees, and higher incomes—and their ability to delay gratification in nursery school. He conducted a full follow-up study in 1988 that clearly demonstrated this correlation. A second follow-up study in 1990 showed a similar correlation in higher SAT scores. When motivational speaker Joachim de Posada re-created the marshmallow experiment with children in Colombia, he was amused when one little girl ingeniously ate just the inside of the marshmallow, hoping to fool the researchers into thinking she had waited. "I predict she will be successful, but we will have to watch her," Posada quipped.

Recently, Stanford psychologist Philip Zimbardo—mastermind of the infamous Stanford Prison Experiment in 1972—also re-created Mischel's original experiment, and met with the same results. Based on this work, he developed the Zimbardo Time Perspective Inventory, a forced-choice quiz that groups people into six different categories based on their "temporal orientation." These categories reflect how our attitudes toward the past, present, and future influence our odds of becoming successful.

Those who are Future Oriented tend to resist the temptation to indulge immediately in exchange for a bigger payoff in the future, just like the kids who patiently waited to eat their marshmallows, while those whom Zimbardo terms Present Hedonistic are like the children who scarfed down the marshmallow right away. They tend to focus on having a good time in the here and now, giving little thought to potential future consequences. Alcoholics score high on Present Hedonism, with a bit of Past Negative thrown in for good measure. (Think De Quincey brooding over all the bad things that happened to him as a child.)

Zimbardo emphasized that we each have varying degrees of all six viewpoints, and the happiest, most successful people are those who achieve a balance between them. Most important, he argues that it is possible to change your time perspective to bring yourself into balance. I took Zimbardo's online quiz and predictably scored highest on Future Oriented. But too much future orientation also exacts a toll: the inability to take time to relax can lead to stress and anxiety and can strain personal relationships. Over the years, I have become much better at indulging my inner Present Hedonist, and I am a happier, more well-adjusted person for it.

What makes one person highly impulsive, and another more able to delay gratification, in the first place? It might be all in our heads. Follow-up brain imaging studies showed key differences between the two groups of children in the marshmallow experiment, notably in the prefrontal cortex, which plays a role in inhibiting impulse. There was more activity in those children most able to delay gratification. Those children who could successfully delay gratification were also far less likely to abuse drugs or alcohol as adults.

In 2010, Vanderbilt researchers used PET scans to examine impulsivity in the brain, and found the heightened response patterns looked suspiciously similar to those typical of addiction—patterns associated with dopamine. Participants were given amphetamines to launch a cascade of dopamine, and then placed in a PET scanner to monitor the effect in the brain. All had been given a battery of psychological tests beforehand, and those who scored highest for impulsivity produced more dopamine in response to the amphetamine "kick." They got a bigger "high" from the drug, and wanted more of it, more often, than those who scored lower on impulsivity.

So just a small difference in brain structure can give rise to dramatic personality differences. And what is the source of such differences? Your genes, of course, specifically a class of proteins called neurexins that glue neuron cells together to keep synaptic connections stable. Proteins are large molecules, so the encoding genes that produce them are also large. In general, the larger the gene, the more likely it is that there will be tiny mutations, some of which could alter the resulting protein in turn. One wouldn't go so far as to say those mutations directly cause impulsivity, but there is intriguing evidence that they might be involved somehow. And subjects who score high on impulsivity are more likely to abuse drugs or alcohol.

There is mounting evidence that the brains of alcoholics are markedly different from those of non-alcoholics. Chronic alcoholics have smaller, lighter, and more shrunken brains, particularly in the hippocampus and that all-important frontal-lobe region that controls planning, decision making, and impulse control. There is less gray matter (neuron cells) and less white matter (axons), which consists of long fibers that are key to relaying information between connected brain cells. The denser the fibers, the more efficient the brain is at processing information. In heavy drinkers, those connections aren't as dense as they are in non-alcoholics.

But how do we know that those differences in the brain weren't present *before* those people started drinking, thereby contributing to the alcoholism? To answer that question, you would need to capture an image of the brain in early adolescence, before someone ever takes a drink, and then track changes over time, as the subject's drinking habits evolve. After all, "Nobody develops an alcohol dependency overnight," said Danielle Dick. "There's usually a trajectory of risk-related behavior."

This Is Your Brain on Booze

If De Quincey has a literary heir apparent, it just might be late night TV talk show host Craig Ferguson, who dished on his struggles with alcoholism in his 2009 memoir, *American on Purpose*. Like De Quincey, Ferguson doesn't stint on the less-than-savory details, although his often-hilarious account serves as more of a cautionary tale to those tempted to follow in his footsteps. You would never accuse him of glamorizing addiction. Unlike De Quincey, Ferguson found his way to bona fide recovery. He memorably pinpoints the moment in his life when he went from binge drinking for fun to a full-blown alcohol dependency: he was in a car with friends one morning, heading home after a night of hard drinking, when he started to feel dizzy. That feeling quickly intensified:

> Immediately I felt a wave, no a tsunami, of absolute terror flood my system. . . . [T]he lack of any apparent cause made it even worse. Clearly I was going mad. I began to feel as if my jaw were not connected to my face. I rubbed at it frantically, twitching and fretting and squirming in my seat. I had broken into a running flop sweat. . . . I knew instinctively that there was only one way to stop the nightmare that was occurring inside of me. For the first time in my life . . . I *needed* a drink.

Ferguson followed a classic developmental trajectory to chronic alcoholism. He started drinking at fourteen, forcing down enough cheap wine laced with rum to get the sought-after buzz, despite the fact that it tasted a lot like cough medicine. Not only did he make himself horribly sick, but he also blacked out, with no recollection of the night's events beyond that point. Blackouts occur because large amounts of alcohol interfere with key receptors in the brain, blocking the neural circuits that create memories, while leaving other basic functions intact. It is unusual for this to happen on one's first drinking spree, however, a fact that, in Ferguson's case, suggests a possible genetic predisposition. Regardless, the experience didn't put him off booze for long; within three months he was drinking regularly. It takes between five to fifteen years of continuous heavy drinking before one develops a clinical dependence. Ferguson transitioned from *wanting* a drink, to *needing* a drink, right on schedule.

That still doesn't explain why Ferguson became an alcoholic while most of the other kids in his neighborhood did not, despite growing up in the same hardscrabble environment and engaging in the same levels of binge drinking. (Some no doubt did, but the majority did not.) Why not his siblings, who grew up in the same home and likely shared similar genetic predispositions? If we could put the fourteen-year-old Ferguson through a brain scan and peek into his brain, what might we find?

Chances are the scan would show less activation in that part of the brain associated with working-memory tasks, and less connectivity between all those crucial white-matter fibers that carry electrical signals between neurons. Adult alcoholics have less dense white matter in key regions of the brain; the roots of those anomalies might already be present in adolescence, before a child ever takes a drink.

Granted, children in general have less white matter when they are young, and brain function is broadly distributed across many different regions as a result. "In younger kids, several regions of the brain are activated to perform many different tasks," explained the University of California–San Diego's Lindsay Squeglia. "It's really inefficient." But as children age, the brain starts compartmentalizing, assigning certain functions to specific regions, thereby becoming more efficient at processing information. You can see this process in the brain's structure: gray matter decreases and white matter increases, becoming denser and thicker to forge stronger connections between neurons. At least, that's what usually happens. In some children, the brain develops a bit differently—and this might make them more vulnerable to alcoholism later in life.

Squeglia and her UCSD colleagues took a group of middle-class, suburban adolescents (ages twelve to fourteen) from the surrounding La Jolla suburbs—a fairly homogenous sample, to help control for other environmental factors—who had never had a drink, and put them through a battery of standard cognitive tests, augmented by an fMRI scan. This established a baseline for each child. Every three months, the researchers would follow up to check on each child's substance use. Eight years later, many of them had begun drinking heavily, while others stayed relatively sober. Although teenagers don't drink as often as adults, when they do, they tend to binge. Squeglia found her boozing teens knocked back an average of thirteen drinks in just one night.

So teenagers like to binge; so what? When Squeglia went back and examined the baseline brain images of those teenagers who went on to drink heavily, she found they predicted a teen's future alcohol use far better than other known risk factors.

Even then, there were indications that their brains weren't processing information as efficiently.

The control group—those who did not become binge drinkers—showed the expected increase in white matter and processing efficiency as they aged, but the opposite occurred in those teenagers who drank, particularly in the frontal lobe regions. This might explain why alcoholics notoriously appear to be stuck at the adolescent phase of psychological development. They still have that "teen brain": the areas associated with reward mature before those regions involved with exercising restraint, so teenagers are especially prone to risky and impulsive behavior. For some reason, chronic alcoholics rarely grow out of this asymmetry between desire and inhibition.*

But let's not discount environmental influences entirely. Danielle Dick has conducted multiple studies of the drinking habits of twins in Finland. Why twins in Finland, you might ask? Finland has a central population registry, enabling Dick and her colleagues to study every twin born in the country over a certain period of time (more than 10,000 and counting). Studying twins tells us something about the relative importance of genetic and environmental influences on a behavior, or how our genes and environment can interact. The home environment does matter in those crucial early formative years, at least when it comes to how early a child starts drinking. Parental influence decreases sharply in adolescence, however, when children's peers become more influential, and any predisposition to alcoholism also becomes a much stronger factor.

---

*There is evidence from neuroimaging studies that these brain differences can be at least partially reversed. The longer a recovering alcoholic stays sober, the more white matter he or she develops in those critical frontal lobes—even more so if he or she doesn't smoke.

Take the GABA-related gene, for example. Dick has found that teens who had that gene were more likely to exhibit the risky behaviors that can lead to drinking problems if they came from homes with "low parental monitoring." Those kids whose parents were "high monitoring"—insisting on keeping tabs on their children's whereabouts, asking for contact information, and setting curfews—were less likely to exhibit those behaviors, despite having the same genotype. "So the environment is clearly helping reduce the expression of that predisposition," said Dick. Even then, parents often discover they can employ the exact same parenting strategies as they did with one child, only to have a sibling turn out completely different. "They say that every parent is an environmentalist until their second child," she joked. "You start to realize children aren't blank slates."

It might be tempting to conclude that we are as helpless in the face of our genetic predispositions as Heberlein's drunken fruit flies. But remember that while genes are deterministic up to a point, they are not our destiny. It is true that an alcoholic has an unfair disadvantage when it comes to exercising self-control; that is why 90 percent experience at least one relapse on the road to sobriety. "We are always blaming people for their own behavior," said Nicholas Grahame, who has a very low tolerance. "When I drink, I feel like crap and just want to go to sleep. But for someone genetically different from me, it feels very different to be intoxicated. The compulsions become overwhelming."

In other words, different people respond differently to alcohol. Some people get a buzz from drinking, while others just get depressed, or even angry. Those who get a buzz from the booze are far more likely to become alcoholics. Despite their higher tolerance, heavy drinkers are actually more sensitive to booze, at least to its euphoric effects; they are less sensitive to its sedative ("downer") effects. Andrea King, who led a 2011 study at the University of Chicago, divided nearly two hundred participants between the ages of twenty-one and thirty-five into groups based on whether they were light or heavy drinkers. All were given one of three drinks: a placebo, a low-alcohol drink, or a high-alcohol drink, flavored to disguise which drink was which. Then the participants filled out a survey after describing their mood, and took a breathalyzer. The light drinkers found the booze made them feel sleepy and sluggish, just like Grahame (and me). The heavy drinkers reported positive moods.

How might this translate into real-world behavior? Craig Ferguson and I both became horribly sick after our first binge-drinking experience. My response was to avoid binge-drinking thereafter; his was to keep drinking heavily until he built up a tolerance. Many factors contributed to our choices, but our very different physiological reactions likely played a critical role.* I became sluggish and sleepy from the alcohol; he found it lifted his spirits. So he had a stronger motivation to keep drinking, despite the ill effects, whereas I was strongly motivated to avoid overindulging.

That is why I am the ultimate Cheap Date. Despite a slightly elevated genetic risk for alcohol dependency, other genes predispose me to low alcohol tolerance, bolstered by additional factors. Chances are, a brain scan would show relatively normal

---

* In December 2012, scientists at King's College, London, identified a gene that seems to be linked to binge drinking in a study of 663 teenage boys. Those with one particular version of the gene showed boosted dopamine levels in response to alcohol and hence experienced a stronger sense of reward from heavy drinking.

adult levels of white matter, particularly in the prefrontal cortex. I don't score high on impulsivity, I am "future-oriented," and I was painfully shy and introverted during those crucial high school and college years. (I once spent an entire junior high dance holed up in a stall in the ladies' room, lest I found myself in the terrifying position of having to make casual conversation.) There were advantages to being socially invisible: I never got invited to the raucous parties where binge drinking was de rigeur—and I would not have much enjoyed them even if I had attended. I also benefited from the educational and economic advantages conferred by growing up in a stable, middle-class home, with "high monitoring" (some might say "controlling") parents. No single factor made much difference on its own, but in aggregate, they all conspired to stomp any risk factors into the ground. And for that I am humbly grateful.

Like Grahame, Heberlein advocates a modicum of compassion for those who struggle with addiction: "We don't want alcoholics to feel so powerless that they spiral into shame and depression." But that doesn't mean she absolves alcoholics of all responsibility for their decisions and actions. While people might feel as if they have no control over such compulsory behaviors, human beings have something her fruit flies lack: that all-important prefrontal cortex, center of decision making and higher cognition. We are never completely at the mercy of our basic instincts—even if some of us have to work a bit harder to exercise that self-control.

5

My So-Called Second Life

> Your avatar can look any way you want it to, up to
> the limitations of your equipment. . . . You can look
> like a gorilla, or a dragon, or a giant talking penis in
> the Metaverse.
>
> —NEAL STEPHENSON, *Snow Crash*

Five minutes before the scheduled event, a few last-minute arrivals trickled into the amphitheater. It was a diverse crowd: men in jeans and casual dress shirts and women in sundresses or workout gear shared space on the benches with heavy-metal rockers, punks with Mohawks, fanboys (and girls) in full steampunk regalia, and the occasional pixie, troll, or elf.

They were all eager to hear author Seth Mnookin chat about vaccines, autism, and denialism, the subject of his book *The Panic Virus* with MIT science writing professor Tom Levenson. As the two men took the stage, Mnookin's clothes suddenly vanished and he stood stark naked at the front of the amphitheater. There was a brief moment of shocked surprise among the onlookers before wry amusement set in. Meanwhile, the organizers scrambled behind the scenes to get Mnookin's clothes back on.

No, this was not Mnookin's worst nightmare, a rehash of that common anxiety dream of showing up naked in class that most of us have had at some point in our lives. The event was taking place in Second Life, a virtual world in which people create personalized avatars and use them to navigate and interact online. There are virtual shops, dance clubs, museums, research groups, churches, and party-friendly beach cabanas, as well as "live" events. This was Mnookin's first experience in Second Life, and he was unfamiliar with the user interface. He simply hit the wrong command, and instead of making his avatar take a seat, he made his clothing disappear. After a few more clicks, the clothes reappeared, and the momentary awkwardness subsided. The rest of the event proceeded without a hitch.

Launched in June 2003 by Linden Labs, Second Life now boasts 15 million users.* The numbers are even more eye-popping when you include other virtual spheres. The British research firm KZero estimates that a staggering 1.8 billion people worldwide have avatars in virtual worlds, and just over 1 billion of those are between the ages of five and fifteen. "Children are growing up immersed in this new paradigm," said Jacquelyn Morie, who studies the effects of immersive technologies and virtual worlds at the University of Southern California's Institute for Creative Technologies. "They will never know a life where they could not put on new avatars just as easily as we put on outfits."

---

*For comparison, in 2008 the hugely popular online game World of Warcraft had 11 million subscribers.

Just what is so appealing to so many people about having a second life? I suspect it has to do with the high degree of control available to "residents" to shape their image in the virtual world precisely to their liking—far more than any of us can exert in the so-called real world. Morie, for one, balks at the distinction between real and virtual worlds, arguing persuasively that Second Life is every bit as real as your offline life. In your first life, you might be a pudgy schlub with a boring job and cramped apartment who drives a battered Toyota, but log into Second Life and you instantly become good-looking, fit, well dressed, and wealthy enough to tool around in a sleek Ferrari—when you're not flying, that is, or instantly teleporting from place to place.

Yes, you can defy the laws of physics. You can even use the built-in "camera" to zoom in and out and rotate your perspective on the virtual space. Perhaps a better question might be why more people don't lose themselves in Second Life. "Some people actually find themselves in Second Life," said Sherry Reson, producer of a Second Life webcast series called *Virtually Speaking Science*.*

Everything in Second Life is user-generated with the three-dimensional modeling tools and scripting language built into the software. There is a bona fide economy, in which residents can exchange "Linden dollars" (the official Second Life currency) for goods and services: clothing, hairstyles, houses, furniture, property, artwork, pets, and specific gestures and animations. Danc-

*In late 2012, I joined Levenson and MSNBC's Alan Boyle as a host for *Virtually Speaking Science,* weekly hour-long conversations with prominent scientists and authors hosted by the Exploratorium in Second Life. The conversations are taped and air simultaneously as a podcast on *Blog Talk Radio.*

ing is one of the most popular activities, followed by having virtual sex. In addition to genitalia, you can purchase a wide range of positions, kisses, embraces, thrusting motions, and "adult accessories." Presumably those residents who engage in these acts derive satisfaction from the experience, even if the user interface is a bit clumsy and awkward. Indeed, for those with physical disabilities, virtual sex may be the only available option, as Reson discovered in discussions with residents from those communities. "It taught me that the opportunities afforded for arousal and release [in Second Life] are life changing and life sustaining," she said.

My own activity in Second Life is relatively mundane. I joined when my spouse—whose dashing avatar is named Seamus Tomorrow—started giving occasional physics lectures as part of the Meta Institute for Computational Astrophysics (MICA), a nonprofit science group that held public events in Second Life until the program ended in 2012. We have a modernist-style, sparsely furnished house, and a virtual feline named Miss Kitty who scampers around with her toy ball and purrs when we pet her. Once, out of curiosity, we used the virtual camera to zoom into the interior of a nearby house, only to find a naked avatar couple copulating enthusiastically: "Ack! Zoom out! Zoom out!" We genuinely felt as though we'd invaded their privacy. That is why the built-in camera has been aptly dubbed the "PervCam."

The reaction to Mnookin's naked avatar, or to stumbling upon our virtual neighbors having a bit of sexy time, is "not just a cold cognitive thing," said Matthew Botvinick, a neuroscientist at Princeton University. "There are all sorts of motivational and social dimensions to self-representation. If something bad happens to your avatar in a social context, then that will engage the same neural circuitry that is engaged when something happens to the actual you in a social context." Several years ago, Botvinick participated in an experiment in which researchers inflicted mild pain on people in an fMRI scanner. The subjects' responses made for a sound study, because the part of the brain activated by pain is well established. The researchers then asked the same participants to view videos of others experiencing pain and found that this activated the same brain regions. The extent to which this occurred depended on how closely they identified with the other person.

Avatars are a virtual extension of the self. Growing evidence suggests not only that we bond psychologically with our avatars, but that those bonds are stronger the more similarities we share with our online selves. As technology continues to advance, one day that bond may become a physical link as well. Technically, an avatar is just an object, and only a pixilated one at that. But the objects with which we surround ourselves can nonetheless telegraph a great deal about who we are.

Tokens and Totems

"Surely you don't believe in that nonsense."

It was intended as a rhetorical question, uttered with an implied wink and a smirk, conveying the unspoken assumption that, as reasonable human beings, we would both agree on this. The speaker, an ardent skeptic who prided himself on his rational approach to life, meant no offense. He was merely surprised to find that I, a lover of science, tote a battered key chain embossed with my astrological sign: Taurus. I've carried it with me for twenty years, like a personal totem.

It was perfectly reasonable for my skeptical inquirer to assume my key chain says something about me. He was employing cue utilization. We all make snap judgments when we meet new people, relying on certain cues to make assessments, and those judgments often can be accurate, at least in broad strokes. Physical attractiveness, race, gender, facial symmetry, skin texture, or facial expressions and body language are all factors that contribute to how we form our impressions of people. Those cues may also include our "stuff": our choices in fashion, jewelry, tattoos, and key chains all provide clues about who we are, whether we intend them to do so or not.

Social psychologist Sam Gosling is very interested in checking out our stuff, but not in a creepy, voyeuristic way. He has studied how we fill our spaces with material things, particularly offices and bedrooms, to better understand what those choices say about our personalities. For instance, certain items function as "conscious identity claims," things we choose based on how we wish to be perceived by others—the posters, artwork, books, or music we display, for example, or the tattoos we ink onto our bodies. We also fill our personal spaces with "feeling regulators": photographs of loved ones, family heirlooms, favorite books, or souvenirs from travel to exotic locales—anything that serves to meet some emotional need. "If you are missing someone, you carry a photo in your wallet, or propped up next to your computer, or you value a necklace that somebody gave to you," Gosling explained. "You do these things to connect to someone as a sort of proxy, until you see that person again." Finally, there is what he terms "unconscious behavioral residue," cues we leave behind in our spaces as a result of our habits and behaviors. A highly conscientious person may alphabetize their books, while the books of some-one who is less conscientious would be more haphazard and disorganized.*

All these cues, taken together, paint a fairly accurate rough sketch of the personality behind them, even those that are not chosen consciously. Gosling's research showed that it is possi-ble to scan the objects in someone's personal space to make in-direct inferences about certain personality traits. Those people who score high on openness to experience on the Big Five per-sonality inventory tend to fill rooms with a greater variety of books and magazines, while those who score high on conscien-tiousness tend to have clean, well-lit, meticulously organized bedrooms.

However, Gosling cautions that this is an imprecise method; we can misread those cues. We may realize a given item is sig-nificant in some way to the owner, but we may not infer cor-rectly the statement that it is making. Context is key. Position can help distinguish whether an object is serving as an identity claim or a feeling regulator. If you walk into someone's office and there is a wedding photo on the desk facing outward, so it can be clearly seen by visitors, that is likely an identity claim. However, if the same photo is turned instead to face the owner, then it likely functions as a feeling regulator, to remind him or her of a loved one.

That is what happened in my encounter with the skeptic who scoffed at my choice of key chain. It does say something about me, but he interpreted it as an identity claim, when in fact it is a feeling regulator. There is a story behind that key chain,

---

*A friend of mine organizes his classical CD collection by birth year of composer; I suspect he would score quite high on conscientiousness.

or rather, a singular person by the name of Nick. We became friends as eager young twentysomethings in New York City, when we both ended up working briefly for the same legal publisher. Nick was smart, sarcastic, and flamboyantly funny, a natural raconteur who could hold a roomful of dinner guests in rapt attention, usually doubled over in laughter while he recounted his decidedly Rabelaisian adventures as a young gay man in Manhattan. He loved good food, good clothes, good music, good sex, and never shied from offering to buy the next round of drinks.

But this was at the height of the AIDS epidemic, when an HIV-positive diagnosis was akin to a death sentence. Over the course of three years, I watched my friend wither away to a shadow of his former self as the virus ravaged his immune system, although his wicked sense of humor and big heart remained intact. Nick even helped me with a last-minute move on a hot summer day after I was evicted from what turned out to be an illegal sublet, stopping every now and then to dramatically wipe the sweat from his brow and announce, "I shouldn't even be doing this, you know. I have a terminal illness!" Then he would grin at my guilty expression and give me a hug to let me know he was just teasing.

Nick's own housing situation was even more precarious: unable to work as his illness progressed, and unable to sign a lease with a reputable landlord, he relied on local city charities and the tight-knit gay community to snag a cluttered, roach-infested grungy basement apartment in Chelsea whose prior occupant had died. It was rent-controlled, and thus dirt cheap, but the stench of death still hung over the place—likely due to the decaying rodent corpses trapped behind the sagging walls. Nick hated it and found it unbearably depressing. Soon he was back in the hospital, and when I went to visit, he broke down in tears over a game of cribbage and begged me to help him find another option: "Please—I just don't want to die there."

Quite frankly, I didn't want to do it. I was barely scraping by myself, and we had only just gotten him settled in that dank basement. It was exasperating to have to start the housing search all over again. But how could I refuse my friend just because it would be inconvenient? It took a good bit of bureaucratic wrangling, but I found him a small sun-filled studio on the East Side. While purchasing a few last-minute bathroom accessories, I spotted a bright metallic red key chain for sale at the register, embossed with Nick's astrological sign: a fellow Taurus. It seemed like the perfect trinket to hold the key to his new apartment, a ray of hope after weeks of darkness.

Eight months later, Nick was dead. I've carried that key chain ever since, the only thing I have left from his long-dismantled life. My totem says nothing about my belief in astrology, although you might be forgiven for making that assumption. To me, it's a token of the beloved friend I lost—the very definition of a feeling regulator—and a constant reminder not to take my friends for granted, no matter how busy I become, because they might be gone sooner than I think. It is also a symbol of the most valuable thing I ever gave to Nick: a decent place to die.

Every personal item has a story behind it, at least if it holds any real meaning for the owner. Cultural historian Mihaly Csikszentmihalyi has argued that we are attached to old photographs, family heirlooms, or seemingly insignificant trinkets precisely because they keep us grounded in the present, and help us remember the past. In that sense, the objects with which we fill our homes play a vital role in how we construct our sense of self. Like Gosling, he lumps such totems into three distinct categories. There are objects that serve as symbols of status, or of good taste. There are objects relating to what he terms "continuity of self" that help construct memory and personality. Finally, there are objects of relationships, like my Taurus key chain, that link us to our loved ones and broader social networks. "Without external props, even our personal identity fades and goes out of focus," he writes. "The self is a fragile construction of the mind."

I might quibble with Csikszentmihalyi's insistence that the self is a fragile construct—on the contrary, the self strikes me as surprisingly robust despite, or perhaps because of, its remarkable fluidity—but his insights into how we infuse material objects with meaning fall right in line with Gosling's research. Gosling found that this phenomenon carries over into our online identities as well: one can infer quite a bit about somebody's personality by perusing his or her Web site, blog, or even an e-mail address. (Many Internet hipsters still sneer at those who use AOL or Hotmail addresses, for example.) We form very different first impressions of someone whose e-mail address is just their first and last name, versus someone who uses the handle "sexyspacekitty69."* Nowhere does this become more apparent than on the social networking site Facebook, where we create detailed personal profiles of our likes and dislikes, share links, play games, take quizzes, and post personal photographs. As of 2011, there were more than 600 million active users in the United States alone. Increasingly, for many of us, our Facebook page is where we keep our stuff. Your Facebook profile is one gigantic identity claim, whether you realize it or not.

---

*I made this e-mail handle up at random, but it would not surprise me to discover that it actually exists.

Gosling drew his conclusions from two related studies. In the first, participants took the Big Five personality test, and those results were compared to the so-called virtual residue (similar to Gosling's behavioral residue in the object study) strewn throughout their respective Facebook profiles. Analysis revealed significant correlations between the self-reported Big Five test results and certain personality traits suggested by the subjects' Facebook profile pages. Extroverts had the most friends and interacted far more frequently than introverts, while those focused, achievement-oriented conscientious types used the site the least. Those with low scores on conscientiousness were far more likely to use Facebook to procrastinate.

You might argue that both the answers to the personality tests and the profile pages were generated by the participants themselves and hence lacked objectivity. So in the second study, nine undergraduate research assistants looked at only the archived Facebook profiles of the study participants and rated their personalities based solely on carefully selected cues: number of photos and photo albums, number of wall posts, group memberships, total number of friends, and even how many words each participant used in the "About Me" section. Once again, there were strong correlations between the profiles and the self-reported assessments: extraversion correlated with the number of friends and higher levels of online engagement, and openness correlated with the number of friends. It proved much more difficult to draw correlations between the cues found on Facebook profiles and the traits of conscientiousness, agreeableness, and neuroticism; the results were inconclusive.

But the two studies aptly demonstrate that your online and offline identities overlap on Facebook, and your profile does reflect your most easily observable personality traits. The same should hold true for Second Life, another online repository for "stuff"—or as Gosling might say, identity indicators. Many residents amass vast personal inventories of virtual items, which may also serve as totems. Reson, for instance, cherishes a virtual rainbow feather—a gift from a friend and mentor, and her very own pixilated feeling regulator.

At the end of the day, Facebook is just one more tool we use for self-verification: we want to be known and understood by others in keeping with how we feel about ourselves. There are two distinct psychological models that we might apply to the social networking site. There is Objective Self-Awareness (OSA), first proposed in 1972, which holds that we are both the subject of our own life stories through our actions and an object when we evaluate ourselves, making reference to societal norms and standards. We can accomplish this simply by looking in a mirror and wondering if those jeans make us look fat. Facebook, too, can serve as a social mirror, whereby we compare our own profiles with those of others in our network. Alas, there is a downside to such behavior. This kind of self-evaluation often results in decreased in self-esteem. We inevitably come across people who are smarter, better-looking, or more successful, and we never quite measure up to those impossible standards. The same is true for our online interactions.

The Hyperpersonal Model might be a better fit. Developed by a communications professor named Joseph Walther in 1996, this model holds that we have more control over how we present ourselves in so-called computer mediated communication than we do in traditional face-to-face interactions. Call it selective self-presentation. When we design our profile pages, we instinctively highlight those features that show us in the most flattering light, thereby creating more positive first impressions. More recent studies found that viewing your own Facebook profile actually boosts your self-esteem—even more so if you view only your profile, rather than comparing it to the profiles of your friends.

In both cases, what we really desire is positive feedback that bolsters our self-esteem and makes us look good, even as we claim to desire objective feedback, in the same way that the subject of a magazine profile claims to desire objective reporting. Perhaps that is why Facebook is so addictive. The very act of updating your profile makes you feel good about yourself, because you are consciously focusing on the positive events and attributes you wish to share with your circle of friends. It's much like staring into a mirror and repeating the mantra of *Saturday Night Live*'s Stuart Smalley: "I'm good enough, I'm smart enough, and doggone it, people like me!"

There is nothing intrinsically wrong with that. Affirmation can be healthy, unless it becomes an obsession, as in the cautionary tale of Snow White's wicked stepmother. The Evil Queen's most prized possession was her magic mirror. Every day, she looked into the mirror and asked the same question: "Mirror, mirror, on the wall, who is the fairest of them all?" And every day, the mirror would reaffirm her unparalleled beauty—until that upstart Snow White came of age, at which point the queen turned into a homicidal maniac. She desperately needed that hyperpersonal self-verification and responded badly to objective self-awareness. Today, the Evil Queen might obsessively log onto her Facebook page to get the daily affirmation she craved. If she were in Second Life, she would obsessively gaze upon her avatar.

I'll Be Your Mirror

In *Harry Potter and the Sorcerer's Stone*, everybody's favorite boy wizard stumbles onto a magical mirror at Hogwarts in which he can see his dead parents standing next to him. But when he excitedly tries to show his best friend Ron Weasley the image, all Ron can see are reflections of himself—or rather, an idealized version of himself, in which he is a star athlete, no longer in the shadow of his numerous older siblings. Harry asks his headmaster, Albus Dumbledore, why they saw different things, and Dumbledore explains that this particular mirror only reflects that which we most desire. Harry longs for his lost family to be whole again, while Ron craves the attention and glory that always seem to elude him. Unlike the Evil Queen's magic mirror, which reflects physical appearance, the Mirror of Erised delves beneath the surface to reflect one's inner self and most passionate desires. If the ongoing research on virtual worlds and self-representation proves correct, our avatars may reveal just as much about us, serving as virtual mirrors.

The word "avatar" comes from a Sanskrit word, *avatāra*, describing various incarnations of the Hindu god Vishnu on Earth. Just as Vishnu inhabits and animates his human host on the terrestrial plane, a person can build a digital representative through which he or she can interact in the virtual sphere. Practically since the invention of games more than four thousand years ago, human beings have employed tokens to represent the players. In the classic board game Monopoly, one may choose from a number of small metal pieces to mark one's place on the board. When people began interacting online, many chose small static thumbnail images to represent themselves, and this practice naturally extended to computer games. The term was first used to describe full virtual bodies in the 1986 online role-playing game Habitat, although Neal Stephenson's ground-breaking 1992 cyberpunk novel, *Snow Crash*, is widely credited with bringing "avatar" into mainstream culture.

Our avatars are the primary means by which we make identity claims in virtual worlds. Almost everything about that design process involves some kind of conscious choice: hair color, hair style, eye color, body type, tattoos, gender, age, ethnicity, clothing style, even our choice of genitalia (should we go that route)—all reflect the user's personal taste and sense of self. Sherry Reson views her avatar as "both a psychological projection and [form of] creative expression [that is] constantly evolving," she said. "The glasses I wear make me feel more like me. And there are times I've dressed as a hermit because it accurately represented my feeling state."

My avatar is named Jen-Luc Piquant. The roots of her name date back to the early days of the Internet. Having a clever e-mail handle was all the rage, and I settled on "lucrezia," in honor of the renowned sixteenth-century femme fatale Lucrezia Borgia, who I always felt got a bad rap as a smart woman in a powerful family who was denied much of that power for herself. She also owned fabulous poison rings, which suited my morbid Gothic sensibilities. I even bought a few cheap knockoff rings from the vendors on St. Mark's Place in Manhattan's East Village. I filled them with salt (instead of arsenic) and would sometimes offer to sprinkle some on a date's food. If he laughed, he was worth a second date.

Fast-forward many years later. I had moved to Washington, D.C., and my upstairs neighbors took to calling me "Jen-Luc"—a mashup of my first name and e-mail handle. When I started my blog in 2006, it just made sense to name the moody little avatar at the start of each post "Jen-Luc." I added the last name "Piquant" as a nod to Captain Jean-Luc Picard of *Star Trek* fame, of course, but also to capture something of the flavor of the blog.

Maybe it's because I named her, but Jen-Luc Piquant took on a life of her own, becoming a full-blown character. She's smart, snarky, and a bit pretentious, sprinkling bons mots throughout her online commentary. Cropped purple hair, a beret, and a black turtleneck are her trademarks—very faux-French— although she has occasionally appeared as a vampire, a ninja, a pirate, a princess, and once as the cartoon character The Tick. She has a penchant for gourmet cuisine, high fashion, celebrity gossip, and existential angst, and she dabbles in both amateur scientific research and Lacanian literary criticism. Over the years she's acquired a middle name—Marie-Evangeliste—and a rumored ex-husband (Crackle of Rice Krispies' Snap! Crackle! Pop! fame) following a wild weekend in Las Vegas, although she claims those are lies, vicious lies. It seemed only natural, when I joined Second Life, to let Jen-Luc into that virtual world as well.

An avatar can be an accurate representation of one's actual self, or a fantasy self (an elf, a dragon, an Amazonian angel), or even an ideal self—the person you might like to be, in a world free of the usual constraints. "We all have selves that we envision in the future, and they serve as very powerful motivational goals," WUSTL's Michael Strube explained. "We may present a self that may not currently be true, but perhaps we would like it to be true. We are hoping that others might accept it and validate it in ways that allow it to become true."

Jacquelyn Morie admitted she had always wanted red hair, green eyes, and fuller lips; in Second Life, she has all three. I designed Jen-Luc Piquant similarly: she is younger and thinner, with fuller lips and better hair—an idealized self that I could never achieve in the meat world, where her proportions would make her a freak of nature. (Apparently she stands six-foot-four; avatars are very tall in Second Life.) That idealization can also apply to personality traits, whereby we create avatars that are more extroverted or less neurotic than we might be in our offline lives.

People occasionally express confusion over whether Jen-Luc Piquant is "really" me, particularly since I use her for my Twitter handle as well. She is more like an alter ego or evil twin. She is far more narcissistic, at least according to the standard Narcissism Personality Inventory (NPI). The NPI consists of forty forced-choice questions asking you to select one of two statements. For instance, "If I ruled the world it would be a better place" versus "The thought of ruling the world frightens the hell out of me."

Dr. Drew Pinsky, co-host of the popular call-in TV show *Loveline*, is also an assistant clinical professor of psychiatry at USC, and he started giving the NPI to celebrity guests who appeared on the show—much to the annoyance of his brash co-host, Adam Carolla. Pinsky found that celebrities score slightly higher (17.84) than average. Woman celebrities were slightly more narcissistic than their male counterparts. Musicians were the least narcissistic. Reality TV stars scored the highest, at 19.45, with female reality TV stars scoring near the top of the charts.*

---

* Carolla was unimpressed with these results: "Who cares? Is this groundbreaking, that celebrities are narcissistic? I mean, this is like you found out Liberace was gay."

As one who sometimes jokes to friends that my Native American name would be Crippled By Self-Doubt, it is not surprising that I scored a troubling 10 on the narcissism scale. Apparently I barely have sufficient self-esteem to function in society. Either that, or I am disinclined to admit to my own self-absorption, even on an anonymous quiz. A certain degree of narcissism is healthy, after all; it's what gets us through the inevitable obstacles and disheartening stumbles in life. But what about Jen-Luc Piquant? Her score almost maxed out the scale, which ranges from 0 to 40. The average American scores around 15.3; Jen-Luc scored a whopping 39. She had a brief moment of uncharacteristic humility, which destroyed her shot at a perfect 40. Even reality TV stars aren't as narcissistic as my faux-French avatar.

For all the freedom in constructing online identities, most people—nearly 90 percent, according to a 2010 study—tend to choose avatars that share similarities with their "real world" selves: the same gender, a similar name, and either resembling their real-life physical self, or personifying an idealized version. A 2007 study of online gamers found that only 4 percent of women chose a male character and only 14 percent of men chose a female character, although adolescents were far more likely to engage in gender swapping as a form of identity play, perhaps because at that age our identity and sense of self are still in flux. For the most part, we use our avatars the same way we use our Facebook profile page: as a means of self-presentation and self-verification via online interactions with others. We bond more strongly with avatars that resemble us, and the more we bond with our avatars, the more enjoyable we will find the virtual experience. We need to be able to look at our avatar and feel, "This is me."

That is not to say you can't have more than one avatar to explore different aspects of your personality. Jacquelyn Morie uses three primary avatars in Second Life, including twins who have distinct personalities and different backstories. She even borrowed her husband's avatar once. To Morie, such role-playing is perfectly natural, even healthy. "Our identity shifts all the time and every day, morphing and evolving based on what we are doing now," she said. "I'm not the same person I was at sixteen and I'm not the same person I was last week." As for my faux-French avatar, Morie opined, "She's part of you, but she's not the totality of you, and she may not even be who you are at the moment."

Perhaps Morie had a point. I took the Big Five test again, this time adopting Jen-Luc Piquant's persona. As expected, there were pronounced differences. She is far more extroverted, and much less agreeable; in fact, she bottomed out on that scale, reflecting her breezy lack of concern for others' needs and feelings. Jen-Luc Piquant is not out to make friends; she does not cringe or apologize or ingratiate, and she speaks her mind. We also differ widely on conscientiousness. I am highly conscientious and goal-oriented, but Jen-Luc is, again, near the bottom of the scale. She loves spontaneity and being in the moment, and tends to be careless and disorganized in her intellectual pursuits.

But there were also striking similarities between Jen-Luc and me. We both scored low on neuroticism (within ten points of each other), and we both ranked high on the openness scale. We both are curious, imaginative, and creative, and like to spice up our daily routine with a hefty dose of variety and new experiences. She may be a conscious creation, but there might be more of me in Jen-Luc Piquant than I consciously intended. That said, in my avatar, these qualities are exaggerated to ridiculous extremes for comic effect; were she real, Jen-Luc would have all the makings of a psychopath.

You might think that in a virtual world where everyone can present an idealized self, appearance would cease to matter, but this is not the case. First impressions matter a great deal, even in Second Life. Much like material objects in rooms and offices, or someone's Facebook page, we can use observations drawn from avatar cues (attractiveness, gender, hairstyle) to form personality impressions.

The findings from the handful of studies published to date offer useful tips on effective cues. Pupil size, viewing angle, and the frequency with which your avatar blinks its eyes are critical to first impressions. Avatars with larger pupils are judged to be more attractive, happier, good-humored, and sympathetic, even though we are not consciously aware of that trait. Frequent eye blinking (sixty blinks per minute) is associated with dishonesty, fearfulness, shyness, and anxiety. Reduce the blink rate to twenty-four blinks per minute, and your avatar will appear more sociable and attractive. Avatars viewed from below are deemed more sociable, self-confident, and attractive, compared to those viewed from above, who are deemed weaker and in need of protection. A full frontal view means that avatar will likely be deemed more trustworthy, open, and sympathetic.

Certain characteristics can also be associated with particular personality traits. Attractive avatars with long, stylish hair are usually seen as extroverted. Male avatars with black hair, or wearing jeans, gray shirts, or long-sleeved shirts are seen as introverted, while female avatars with blond hair wearing pink shirts, necklaces, bathing suits, or high heels are deemed more extroverted. Large breasts on female avatars serve as a cue for extroversion, too, as well as openness, although if they also favor Gothic-style clothing, they are seen as more neurotic. Blond hair and dressy clothes on females correlate with higher agree-ableness. Male avatars should avoid army pants, black shirts, and sunglasses, lest they be deemed less agreeable.

That's what the studies say, anyway, which doesn't bode well for Jen-Luc Piquant's social prospects in Second Life. True, she has the requisite flowing long hair, but it is purple, not blond. Rather than enhance her feminine pulchritude to exag-gerated extremes—as is the custom for female avatars in virtual worlds—I reduced her breast and hip size as much as possible to give her a lanky gamine build. She has only two outfits: a snazzy steampunk ensemble, complete with military-style trench coat and chic aviator goggles, and a classic edgy "rocker chick" ensemble. Jen-Luc doesn't do ultra-high heels, French maid outfits, sexy bathing suits, or any shade of pink. As a commit-ted Lacanian, she eschews the most obvious forms of self-objectification, even as she acknowledges her own dual existence straddling the boundary between Subject and Object.

That doesn't stop her from mugging shamelessly for Second Life's built-in camera every chance she gets. The ability to shift perspective means we can view our avatar from the front, and see our virtual self the way other residents see us in cyberspace. We can also take snapshots of our avatar, and when the camera clicks, said avatar will momentarily throw up its hands to frame its face and beam with delight. Morie noted that while many of us react poorly to photographs of our actual selves, "We are more likely to be enamored of the look of our avatar." Second Life is a true digital mirror in that regard. As Dumbledore tells Harry Potter, "The happiest man in the world looks in the mir-ror and only sees himself exactly as he is."

Your Inner Lobster

Neal Stephenson introduced science-fiction fans to the Meta-verse in *Snow Crash*: a fully immersive virtual world that offers a glimpse of what online spaces like Second Life might one day become. In this dystopian future, people escape their grim exis-tence in an anarcho-capitalist Los Angeles by logging into the Metaverse, where they can lead parallel virtual lives through their avatars. This world is threatened by a new virtual drug called snow crash, a computer virus that not only affects one's avatar in the Metaverse, but also the actual brains of hackers in reality, dissolving the boundary between "real" and "virtual," "self" and "avatar."

Stanford University's Jeremy Bailenson has demonstrated similar boundary effects in his Virtual Interaction Lab. One of the first simulations he created was a virtual gaping pit in the middle of a simulated "room" with a board laid across it. Test subjects, outfitted in full VR gear, were instructed to walk across the pit. Even though they knew consciously that the pit wasn't real—they'd seen the real-world version of that room and there was no pit—they still found themselves reacting as if the pit were really there. Some teetered uncertainly, some fell down, some ran away, some screamed in fear. The psychologi-cal responses were very real, a testament to the power of digital illusions.

Inhabiting a virtual world can affect our behavior offline, too. Bailenson devised an experiment to investigate what happens when we view our own digital doppelgängers. He found that watching your digital avatar running on a treadmill, for example, makes you more likely to exercise offline as well.

The effect is even stronger when you watch your avatar become thinner or heavier in response to behavioral choices, such as eating carrots versus candy, or exercising versus standing still. The more we identify with our avatars, the more strongly we will respond. Spend enough time with an avatar that looks like us, and the lines between our real and virtual identities begin to blur. It takes only twenty minutes of exposure to produce changes in behavior.

If your avatar has wings, and you become accustomed to manipulating their movement in cyberspace, your brain might become unable to distinguish between the virtual wings and your "real" body. Virtual-reality guru Jaron Lanier first speculated about this in the 1980s, with his notion of "homuncular flexibility." The homunculus is the brain's map of the body that resides in the cortex, with those body parts requiring the most synaptic connections enlarged with respect to less connected parts. As Lanier recalled, back then, he and his cohorts donned full-body suits covered in sensors to create "bodies" in virtual space. There was the occasional bug, one of which "caused my hand to become enormous, like a web of flying skyscrapers." The glitch made him realize how quickly he learned to adapt to the new body part. And that led him to wonder just how much he could distort his body before his brain could no longer adapt.

His favorite experiment involved a virtual lobster, which had three midriff arms on either side of its body—something the human body lacks, so Lanier wondered how one might learn to control those arms in virtual space. He found that he could mix and match small twists and flexions of his existing human limbs to control the extra limbs on the virtual lobster. Theoretically, the brain should be able to incorporate certain physical attributes of our avatar into its map of the self—even if our avatar has wings and our physical self does not. Lanier is now collaborating with Bailenson on demonstrating this in the lab.

Time is a critical factor in bonding with an avatar, according to Caltech neuroscientist Christof Koch. There must be a sense of continuity, which translates into less than a 250-millisecond delay between the brain sending a command for motion and the feedback it receives once the action is performed. "If the delay is too long between when I initiate the reaction and when the feedback comes, my brain can't really deal with it, and you don't get this easy merging of nerve and muscle," said Koch. "But there is no reason why the brain shouldn't adapt [to an avatar] as long as the delay is relatively short."

Our brain already does this with extensions of our physical body. In the 1930s, German philosopher Martin Heidegger proposed the concept of "ready-to-hand." He reasoned that since we don't consciously think about our fingers while tying our shoelaces, or about our hands while hammering in a nail, in some sense we "fuse" with our most familiar, functional tools. They become part of us, much like Lanier's virtual extra lobster limbs. This is equally true for a blind man who uses a cane to sense and navigate his environment; the cane becomes an extension of his physical body, at least as far as his brain is concerned. It is also true when we use a computer mouse, according to Anthony Chemero, a cognitive scientist at the University of Cincinnati, who performed the first direct test of Heidegger's concept.

Chemero set up a simple experiment in which participants used a mouse to control a cursor on the monitor. The mouse was rigged to malfunction halfway through the test, such that the cursor on the monitor lagged significantly behind the movement of the mouse. You can imagine how frustrating the participants found this malfunction; profanity was a common reaction. Chemero tracked their hand movements throughout the experiment, and when he analyzed the data, he found markedly different mathematical patterns produced when the mouse was functioning versus when it was malfunctioning.

The hand movements when the mouse was functioning fit a pattern known as "pink noise," which appears whenever something is naturally attuned to our cognitive processes. Pink noise is similar to white noise (the snowy static on a TV screen) in that it contains every frequency within the range of human hearing (between 150 Hz to 8,000 kHz), but unlike white noise, it is not completely random. Rather, pink noise is indicative of a system that, while momentarily stable, is literally teetering on the edge of chaos—a sweet spot nestled between rigid order and disarray. Pink-noise patterns can be found in pulsing quasars, heartbeats, the structure of DNA, the flow of traffic, most musical melodies and electronic devices, tides, and the fluctuations of the stock market.

This is a fragile state. Chemero found that all it takes is one small malfunction in the mouse to break the connection and push the pattern over the brink into a new chaotic state. When the mouse was functioning properly, the pink-noise pattern emerged and the mouse was "ready to hand." But the pattern vanished when the mouse malfunctioned. The users were no longer "fused" with the mouse; it was no longer part of their cognition.

Chemero also adapted the experiment to measure physical indicators of stress, such as heart rate, respiration rate, and galvanic skin response (changes in the electrical activity of the skin triggered by emotional or physiological responses). He found an increase in all three at precisely the same moment when the mathematical pattern transitioned from pink noise to chaos. So Heidegger was correct. "You're so tightly coupled to the tools you use that they're literally part of you as a thinking, behaving thing," Chemero told *Wired* in 2010. The brain's unusual ability to incorporate nearby bits of the environment into its concept of self "has to be foundational in any feeling of oneness with your avatar in Second Life," Chemero explained.

It is even possible to fool the brain with a rubber hand. While still a graduate student, Botvinick designed an experiment in which a participant's hand was hidden and replaced by a rubber hand in the position where the real hand would have been. Both the real and fake hands were stroked simultaneously, and even though participants were in on the "trick," and knew the rubber hand was a fake, they still responded as if it were part of their body. Threaten the rubber hand by attempting to stab it with a dagger, for instance, and the participants would exhibit an involuntary startle or fear response. It's the combination of visual and tactile feedback that does it, and it takes only a few seconds for the illusion to kick in. Not only that, but subsequent experiments proved this wasn't a purely psychological effect. The real, hidden hand's temperature actually dropped half a degree—a small but measurable physiological response.

Henrik Ehrsson of the Karolinska Institute in Sweden has taken the rubber-hand illusion one step further and used similar methods to induce out-of-body experiences in subjects, armed with little more than a video camera, goggles, and two sticks. In one experiment, he manipulated study participants in such a way that they felt as though their bodies were of different sizes, either the size of a doll or a giant. Subjects would lie down on a bed wearing a head-mounted display connected to two video cameras. Both cameras faced a fake body lying on the bed next to the subject, so when the subject looked down at their bodies, they "saw" the fake body instead. To get the subjects to "bond" with those fake bodies, Ehrsson combined the visual feedback with tactile feedback, poking the arm or stomach of the mannequin while simultaneously doing the same to the subject. A few seconds was all it took to change most subjects' perception of their physical world.

"Our experience of self is surprisingly malleable," said Matthew Botvinick. He thinks it should be possible to extend this same kind of linkage to one's avatar in a virtual world—at least in principle. "At a figurative level we are constantly putting avatars out there, representing ourselves to people," he said. "The boundary, in my mind, between what we have established coarsely and what we really don't understand corresponds to the boundary between the bodily self versus other forms of self-representation." We would need to take the self that we present to others, find a way to "detach" it via a separate digital avatar, and then create sufficient levels of sensory feedback to attain that all-important coupling between user and avatar.

We don't yet have all the requisite technology to achieve this, but several rudimentary pieces are already in place. The biggest challenge is the coupling between user and avatar. Most existing user interfaces for virtual worlds are as clumsy and nonintuitive as the one in Second Life, limiting the extent to which our avatars can be assimilated into our cognition, thereby becoming truly "ready to hand." Bionavigation systems like Wii or Microsoft's Kinect are more intuitive, making it possible to track physical movements so one's avatar can mimic them, with no need for sophisticated technologies like motion-capture suits.

Jacquelyn Morie, for one, finds this approach equally unsatisfying, especially in Second Life, where the ability to fly or teleport is a big part of the appeal. How do you mimic those physically impossible movements with a Kinect? Users might discover some way to adapt using micromovements, much like Lanier did with his virtual lobster, but Morie envisions something even more radical. Ideally, she would like to design an interface that could detect emotional states in the brain and trigger the appropriate responses in an avatar within Second Life—almost a form of virtual mind-reading.

That kind of nonverbal communication and control would require a brain-computer interface (BCI) of some kind to translate the electrical impulses in the scalp generated by cognitive activity into commands to control a computer cursor. The technology is still in its infancy, although researchers at Duke University successfully trained two monkeys with BCI implants to use their brains to move the hand of an avatar and successfully identify the texture of virtual objects—something they achieved without moving any part of their real bodies, just manipulating their virtual "hands" over the surface of the virtual object. And in 2011, Adam Wilson, a member of the University of Wisconsin's Neural Interfaces Lab, used a BCI to post simple messages to Twitter ("USING EEG TO SEND TWEET" and "SPELLING WITH MY BRAIN"), although that mind control clearly did not extend to disabling the Caps Lock key.

The current techniques used to connect a brain with a computer require attaching plastic electrodes to the body, along with messy conductive gels, as well as hardwire links to circuit boards and bulky power supplies, all of which make it difficult to use BCIs outside a controlled laboratory setting. However, Todd Coleman, a bioengineer at the University of California–San Diego, has devised an intriguing alternative. Coleman's version mounts electronic components onto a thin sheet of plastic covered with a water-soluble layer that sticks to skin. Then the plastic dissolves, so the electronics are imprinted into the skin, much like a temporary tattoo, and can detect electrical signals from the brain and transmit them wirelessly to computer. Even better, the device mimics the stretchy, flexible properties of skin, allowing for natural movement. With such a system, one could be truly "jacked in" to a virtual world, with a physiological link to one's avatar.

Jeremy Bailenson cautions that for all their potential, BCIs are unlikely to be a viable commercial technology in the short term. Creating a realistic digital self, however, might be attainable within just a few years. If you are willing to fork over $400, companies like Second Skin Labs will digitally scan photographs of your face to ensure your Second Life avatar more closely resembles you. ConAgra, owner of the Orville Redenbacher brand of popcorn, dug through years of archival sound clips, video, and photographic footage to build a digital version of Redenbacher, who died in 1995. The faux Redenbacher made his advertising debut in 2006, marveling at the storage capacity of an MP3 player.

Then there is Contour, a camera system that creates realistic synthetic actors by capturing the intricacies of facial movement at resolutions as high as 200,000 pixels. It was used to create a digital likeness of actor Brad Pitt for *The Curious Case of Benjamin Button*, in which the title character ages in reverse. The program re-created older and younger versions of Pitt in such exquisite detail, "he basically never has to act again," said Bailenson, adding that it isn't possible to re-create that same level of detail using just one's current digital footprint. But with the full cooperation of the subject, combined with full body scans, voice matrices, and psychological questionnaires, it should be possible to achieve the equivalent of "total personality downloads" with immersive digital technology—or at least the illusion thereof. It won't be "you" in the sense of a conscious being, but it will be as close as modern science can get to a perfect representation of all that you are, with the added element of enabling others to interact with this digital self. Think of the portraits of deceased headmasters in Dumbledore's quarters at Hogwarts. They capture the personalities of the individuals, and those likenesses even react to the events they witness, but they are still a pale reflection of the people on whom they are based.

Jacquelyn Morie thinks such digital selves will become the snapshots of the future. "My great-grandchildren can come and talk to me," she said. "I may look like an eight-bit video game to them, but it will be charming, like looking at a black-and-white photo, except it will be interactive. They'll have some sense of who I was, not just what I looked like." The first step is to forge that critical connection between brain and avatar. Ideally, an avatar of the future would also learn from being connected, so it could act as a surrogate when the user was offline. It would require not just recording a user's real-life memories, but the ability to create its own memories of its experiences and encounters in the virtual world as well.

This is not the equivalent of the "singularity," a term coined by futurist Ray Kurzweil to describe a future in which everyone would be able to upload their consciousness into cyberspace, thereby achieving a form of immortality. Many neuroscientists remain highly skeptical about the likelihood of ever achieving such a feat; we have yet to map a full connectome, and even that would be insufficient to re-create human consciousness. Morie insisted that, even if it could be done, she would find the experience unfulfilling without a physical body. "Our brain is entangled with this body," she said. Avatars offer a handy repository for our digital selves, however, into which we can download our memories, thoughts, and experiences. She envisions a future not of singularity, but of multiplicity—many different representations of our selves that live on in virtual space: "This is when our multiple avatar representations have become so thoroughly us, and we them, that our essence remains in their crucibles after our deaths."

It's as close as most of us are likely to get to immortality.

6

Born This Way

Aunt Alexandra was fanatical on the subject of my attire. I could not possibly hope to be a lady if I wore breeches. When I said I could do nothing in a dress, she said I wasn't supposed to do things that required pants.

—HARPER LEE, *To Kill a Mockingbird*

In 1911, a locked box was recovered from the manor house of a British nobleman, marked OLD PAPERS—NO VALUE. But you can't always judge a box's contents by its signage. Inside was a rare thirteenth-century manuscript by one Heldris of Cornwall, a fictional allegory about a young woman called Silence who was raised as a boy so that she could inherit her father's estate.

Silence excelled at hunting, riding, wrestling, and armed combat, unaware that she was a girl—until she hit puberty. Once the situation was explained to her, she agreed to keep up the masquerade; she liked her freedom and active lifestyle and wasn't keen on learning sewing and needlepoint. The convoluted plot eventually deposited her at the king's court, where she rebuffed the advances of the lusty Queen Eufeme and was banished to France for her trouble. It all ended happily, with the evil queen's treachery unmasked, along with Silence's true nature, and the king took her as his new queen.

I can relate to Silence's lifestyle. I preferred jeans to dresses as a child, and loved exploring the undeveloped wooded areas around our suburban neighborhood. My horrified mother once caught me crawling out of a dry sewer pipe designed to handle runoff during heavy rains—or, as I preferred to think of it, a "secret passageway." I frequently chose male roles when play-acting with friends. One of my favorites was the Scarlet Pimpernel, because I could engage in mock fencing duels, using a long stick as a foil, while tossing off witty ripostes. When swimming at the community pool, I pretended to be Marine Boy, an animated cartoon character who conversed with fish and chewed a special "aquagum" so he could breathe underwater. And I longed to be allowed to mow the lawn, a chore reserved for my brother. My father finally relented, but the mower proved a bit too unwieldy. He walked around the yard afterward, shaking his head in dismay at the bald patches where the mower blades had cut too deep. At least he let me try.

There was a time, in prior centuries, when both Silence and I might have been deemed "unnatural" in our dress and behavior, unless we were fortunate enough to be born into aristocracy, where such peccadilloes were tolerated, if not fully embraced. (The French mathematician Émilie du Châtelet and novelist George Sand were both notorious for dressing in men's clothing and engaging in unwomanly intellectual pursuits.) Times have changed and it is now much more socially acceptable for a little girl to be a tomboy.

The same cannot be said for little boys who show a fondness for dolls or dresses. A 2012 *New York Times* article featured a boy named Alex who loved soccer and princesses, superheroes and ballerinas, unicorns and dinosaurs. Some days he liked to wear a dress, or paint his fingernails; other days he played with trucks and proudly wore his Spider-Man shirt. Alex is in good company: between 2 and 7 percent of boys under the age of twelve exhibit this kind of cross-gender behavior. These "pink boys" frequently endure mockery for their "sissy" preferences— despite the fact that it is no longer unusual to see grown men with long hair, pierced ears, or manicured hands—and feel pressured to conform to society's expectations. Small wonder that most outgrow this phase by adolescence, or at least learn to squelch their unconventional side.

Why do so many people find gender-atypical behaviors distressing? "People rely on gender to help understand the world, to make order out of chaos," Jean Malpas, who heads the Gender and Family Project at Manhattan's Ackerman Institute, told the *New York Times.* "The social categories of man/woman, boy/girl are fundamental, and when an individual challenges that by blurring the lines, it's very disorienting. It's as if they're questioning the laws of gravity." There is also a powerful undercurrent of fear that such children will grow up to be gay, lesbian, bisexual, or transgender, or otherwise skew from socially accepted norms of gender and sexuality.

This is a messy, confusing area where behavior entangles with personal identity. Few traits are as central to our sense of self as those relating to gender and sexuality—or as fraught with intense emotion. "Gender and sexuality are given so much importance in Western culture, so our personal identities are often very bound up in them," explained Meg Barker, a psychologist at the Open University in England.

Many people mistakenly conflate several different concepts.

Gender expression is distinct from sexual orientation, gender identity, or even biological sex. Sexual orientation concerns erotic attraction; gender identity concerns our internal sense of self, which may or not match up with our biological sex. As the saying goes, "Sexual orientation is who you want to go to bed *with*; gender identity is who you want to go to bed *as*." Gender expression is how we choose to express our gender identity, encompassing clothing, hairstyles, mannerisms, speech patterns, play behavior, even the roles we take on in social interactions.

Sexual orientation is particularly complicated, in part because nobody is quite sure how to define it. For some, it is a purely physiological designation; our sexual orientation can be determined by what (or whom) produces physical sexual arousal. For others, it is determined by our choice of sexual partners. Still others might insist that it's not our actual behavior but whether we fantasize about men or women that determines our sexual orientation. All of these are distinct from how we might choose to self-identify. It is not inconceivable for someone to identify as straight and yet engage in a spot of homosexual behavior now and then, or to fantasize about same-sex encounters.

In the midst of all the action and political intrigue in Heldris of Cornwall's text, the personifications of Nature and Nurture (Noreture) take the time to debate which of them is the true "author" of Silence's identity, six hundred years before Francis Galton coined the terms.* The debate is still raging over whether people choose their sexual orientation, or whether homosexu-

---

*Some historians have speculated that Heldris of Cornwall was also a woman masquerading as a man—medieval England was not a society known for its progressive attitudes toward women—although women disguising themselves as men is a common literary trope.

als, heterosexuals, and everyone in between are simply "born that way." The scientific consensus is still very much in flux, with many controversial findings. Those opposed to the very notion of homosexuality maintain that being gay is a choice (and, for many, a sin), and therefore can be "cured." Members of the gay and lesbian community assert the opposite: they did not choose to be gay and could not change their sexual orientation any more than a leopard could change its spots.

We live in a heterocentric society with rigid binary notions about male and female, gay and straight. Various estimates peg the percentage of the homosexual population at just 3 percent to 5 percent among men, and 1 percent to 2 percent among women. So many people in this community have good reason to fear that any implication that sexual orientation involves choice could be used as political ammunition to deny them basic civil rights. The same is true for those who object to nature-based arguments when it comes to perceived gender differences in men and women, for fear it will be used to rationalize continued gender bias in our culture at large.

These concerns are valid—I share them—but they are not sufficient reason to summarily dismiss controversial scientific findings that don't fit our preferred narrative or political agenda, or elicit intense emotional discomfort. While any study will have its flaws and limitations, and the science continues to evolve, many of the same methods used to study sexuality and gender have also been used to study personality traits, alcoholism, and other complex behaviors, and the same fundamental principles apply.

There is solid evidence that genes play a role in sexual orientation. If one brother in a pair of identical twins is gay, there is a greater chance that his sibling will also be gay; that likelihood drops by half for fraternal twins, although it is still a much higher percentage than in the general population.* If one identical twin sister is gay, there is a greater chance her sibling will also be gay. There also seems to be a genetic component to gender expression. In 2006, Dutch researchers examined twins between the ages of seven and ten and concluded that as much as 70 percent of gender-atypical behavior in girls and boys could stem from genetic factors. Either way, genes matter; we're just quibbling over degree.

However, insisting that sexuality and gender are *purely* genetic with no room for social and cultural influences, personal choice, or flexibility, doesn't fit the data either. During a panel at the 2012 conference of the Society for Neuroscience, University of Cambridge psychologist Melissa Hines cautioned against embracing any explanation that relied on a single factor. It is never just genes, or hormones, or socialization, or cultural artifacts, or unique experiences. "It's impossible to say which is nature and which is nurture," agreed Meredith Chivers, a psychologist at Queen's University in Ontario. "They are constantly working with each other."

Au Naturel

Butterflies seem like such innocent creatures, flitting gaily from roost to roost with their bright, iridescent wings flashing in the

---

*The specific probabilities vary from study to study because there are so many confounding factors. While prior studies found rates as high as 52 percent for men and 48 percent for women, the most recent numbers from a 2000 survey of 5,000 Australian twins showed much lower rates: 20 percent for men and 24 percent for women.

sunlight. So imagine the horror of lepidopterist W. J. Tennent when, while diligently tracking Mazarine Blue butterflies in Morocco in 1987, he spotted several males of the species indulging in mating behavior—not with female butterflies, but with each other. Yes. Those Mazarine Blue butterflies were gay. A shocked Tennent published his observations in the *Entomologist's Record and Journal of Variation,* with this unintentionally hilarious bit of moral harrumphing:

> It is a sad sign of our times that the National newspapers are all too often packed with the lurid details of declining moral standards and of horrific sexual offences committed by our fellow Homo sapiens; perhaps it is also a sign of the times that the entomological literature appears of late to be heading in a similar direction.

One man's "horrific sexual offence" is another man's private ecstasy, and the same goes for the butterflies, oblivious to Tennent's offended sensibilities. Neither are butterflies the only species to engage in same-sex mating behaviors. A British naturalist named George Murray Levick traveled to Antarctica with the 1910–1913 Scott Expedition, and spent several months studying the breeding habits of a colony of Adélie penguins at Cape Adare. Levick was horrified to witness not just male penguins mating with other males, but one young male Adélie penguin attempting to copulate with a dead female. He dutifully recorded this "astonishing depravity" in his notebook, although he did so in Greek so that only educated gentlemen would be able to read it. He wrote a separate short paper in English on the sexual behavior of Adélie penguins, but it was solely for private circulation among experts, since it was deemed too shocking for publication.

Same-sex pairings have been recorded in some 450 different species, from flamingoes and bison to warthogs, beetles, and guppies. Female koalas sometimes mount other females, while male Amazon River dolphins have been known to penetrate each other's blowholes. The observation of female-female pairs among Laysan albatrosses made national headlines, prompting comedian Stephen Colbert to warn satirically that "albatresbians" were threatening American family values with their "Sappho-avian agenda." Female hedgehogs may hump one another or perform cunnilingus, while 60 percent of all sexual activity among bonobos takes place between two or more females. A male penguin couple at the Central Park Zoo raised a chick together, inspiring the controversial children's book *And Tango Makes Three*. One biologist confessed in his memoir to cringing at the memory of watching two male rams mount each other repeatedly: "To think of those magnificent beasts as queers—oh God!"

In fairness, all these "gay" animals aren't really "queer" in the human sense of the word, despite our tendency to anthropomorphize them. Canadian biologist and linguist Bruce Bagemihl prefers to call this sort of thing "biological exuberance," and has claimed that even today, respected scientists are guilty of a certain heterosexist bias. They are so steeped in cultural notions of morality that they can't help but mimic Levick's Edwardian prudery, describing these acts in judgmental terms: "aberrant," "abnormal," or "unnatural." The latter is particularly ironic, given just how common same-sex pairings appear to be in the animal kingdom.

Sexual behavior in animals isn't necessarily indicative of

sexual orientation. Drunken male fruit flies become less discriminating in their mating preferences, often chasing after other males instead of females. The alcohol merely alters their behavior temporarily. In 2007, University of Illinois biologist David Featherstone was researching new drug treatments for Lou Gehrig's disease when he stumbled upon an unusual protein mutation in the brains of male fruit flies. It altered their sense of smell sufficiently that they were attracted to male pheromones—scent-related chemicals key to sexual arousal in animals—in addition to female ones, and mounted other male fruit flies when they tried to mate. Featherstone found he could use this mutation to control the mating behavior of his male fruit flies, making them "gay" or "straight," accordingly, at least in terms of their behavior, although technically, his mutated flies would be bisexual.

So might it be possible to "cure" someone of homosexuality by flipping a genetic switch? That certainly wasn't the affable Featherstone's intention when he stumbled on his protein mutation. He was flabbergasted by the volatile public reaction to his research, with some decrying him as a modern-day Dr. Mengele while others wrote heartbreaking letters begging him to cure their suicidal son or daughter of the "disease" of homosexuality. Even if he had wanted to find such a cure, it is not that simple. Human beings are far more complicated than fruit flies or mice. There is no single genetic switch that makes a human being gay or straight, despite the fervor that accompanied the 1993 discovery of a "gay gene" by Dean Hamer.

Hamer himself never referred to his discovery by that name. He merely identified one region near the tip of the X chromosome in men: XQ28. There are hundreds of genes that make up XQ28, but Hamer found that gay brothers shared genes in this region at much higher rates than gay men shared with their straight brothers. Subsequent studies have been inconclusive: some verified Hamer's findings, while others found no such link. It appears that while XQ28 is involved somehow in the sexual orientation of men, it does not act alone.

One way genes may influence sexual orientation is in the development of the brain. The prudish biologist who mourned the sight of queer rams in his memoir might be surprised to learn that approximately 8 percent of rams engage in homosexual mating habits. That behavior has been linked to a brain region called the ovine Sexually Dimorphic Nucleus (oSDN), which is half the size in "gay rams" as in strictly heterosexual rams. This suggested a possible human equivalent, and in 1991, neuroscientist Simon LeVay announced he had measured a key difference between the brains of gay and straight men, based on autopsies of forty brains. Specifically, he found a small clump of neurons in the anterior hypothalamus—an area linked to the control of sexual behavior—that was, on average, more than twice the size in straight men than in gay men.

LeVay dubbed this clump of neurons INAH3, and subsequent studies verified that INAH3 is generally larger in straight men than in gay men, although this difference has yet to be linked directly to sexual orientation. Some people mistakenly consider this the "gay center" of the brain, but the brain's functions are distributed across many different, interconnected networks. There is no such thing as a "gay region" of the brain, just as there is no single "gay gene."

Other differences have been observed. The corpus callosum, a long fiber tract connecting the two brain hemispheres, is larger in gay men than in straight men. This region and the amygdala develop very early, suggesting they are genetically determined, although it is also possible that the differences may be a result of homosexual activity rather than a cause. In heterosexual women, the two halves of the brain are about the same size; in heterosexual men, the right hemisphere is slightly larger. Gay men's hemispheres are relatively symmetrical, like those of straight women, while gay women's brains are more asymmetrical, like those of straight men. Straight and gay men respond differently to male and female pheromones. Asked to sniff one compound derived from women's urine, and another derived from male sweat, the hypothalamus in straight men lit up when they smelled the female urine compound. The gay men's hypothalamus lit up when they sniffed the male sweat compound—the same way straight women responded.

Scientists are still puzzling over what these differences might mean, but there is evidence that sexual orientation—at least in men—could be linked with prenatal differences that develop in the brain, most likely the result of variations in exposure to sex-specific hormones. It is the hormones circulating in the womb early in fetal development that help determine whether a baby will be male or female, with all the relevant genitalia, internal reproductive organs, and secondary sex characteristics that entails, including breasts, fat distribution, and facial or body hair (or lack thereof). Variations in hormonal exposure may also influence sexual orientation. Testosterone affects developing nerve cells in the fetal brain, steering it in the male direction, or in the female direction if the critical surge of hormones doesn't take place. Varying levels of the hormone might account for the observed differences in brain structure between gay and straight men.

This is the thinking behind the fraternal birth-order effect, whereby the chances of a boy being gay increase with each additional older brother he has in the family. A man with three older brothers would be three times more likely to be gay than a man with no older brothers. The hypothesis is that, with each successive son, the mother produces more "anti-male" antibodies and these block the full "masculinization" of the fetal brain by binding to male-specific molecules on brain cells, thereby rendering them inactive. Even then there would still be more than a 90 percent chance such a man would be straight; it's a very small effect, much more rare than the prevalence of homosexuality in the general population. A man would need to have eleven older brothers to reach a fifty-fifty chance of being gay.

Much of this work has been heavily criticized, most recently by Barnard College sociomedical scientist Rebecca Jordan-Young, who spent thirteen years poring over the accumulated literature and analyzing the methods, statistical practices, and assumptions underlying all that research. On the issue of fetal hormones, Jordan-Young acknowledges in her book *Brainstorm* that they must influence the brain in some way, but adds, "The evidence for hormonal sex differentiation of the human brain better resembles a hodge-podge pile than a solid structure." As for the measured differences in the brains of gay men, straight men, and straight women, she finds those studies too limited in their approach. Human beings don't fit so neatly into those categories. If the animal kingdom tells us anything, it's that nature abhors the binary model for sexual behavior, ditching a strict either/or construct in favor of a richly varied continuum.

Bilateral Shifts

"I've been straight and I've been gay, and gay is better."

With those eleven words, actress Cynthia Nixon launched a media firestorm in 2012, inadvertently angering members of the gay and lesbian community because she implied that her sexual orientation was a choice. To have a prominent member of their own community give what was perceived as ammunition to the opposition was upsetting to many, despite Nixon's many years as a staunch advocate of LGBT civil rights. "There is this fundamental understanding of a core essence of somebody's person, and when the core changes, there is this sense of betrayal," said Meredith Chivers. "That's part of how people have conceptualized sexual orientation as being this stable, relatively inflexible attribute of an individual. People just don't like it when they can't predict other people's behavior."

Initially, a feisty Nixon doubled down, declaring to the *New York Times Magazine*, "For me it's a choice, and you don't get to define my gayness for me." She later clarified her position to *The Advocate*, self-identifying as bisexual, adding, "I believe bisexuality is not a choice, it's a fact. What I have 'chosen' is to be in a gay relationship." The *Sex and the City* star was in a heterosexual relationship with a photographer for fifteen years, a man who fathered her first two children. As Nixon tells it, when she became involved with education activist Christine Marinoni in 2004, it wasn't because she had changed in some fundamental way. She just happened to fall in love with a woman, and chose to act on those feelings. Her sexual orientation was fluid enough to accommodate this new love affair.

Talking about choice doesn't undermine the fact that sexual orientation may be innate. This is not the same as insisting that it is entirely determined by our genes. A trait can be innate and still be the result of a complex interplay between genes and environment. Nixon, as a bisexual, has the luxury of choosing to partner with a man or a woman (or both). But she can't change her fundamental bisexual orientation, no matter how strenuously advocates of gay conversion therapy insist that she can. A 2009 review by the American Psychological Association found that not only is there zero scientific evidence that gay conversion therapy works, it might do more harm than good, making patients more anxious or depressed, even suicidal.

That said, insisting that sexual orientation involves no room for personal choice or flexibility is overly restrictive. "It is far more accurate to say, 'I was born this way and there's some flexibility here, and I can express my sexual orientation in several different ways,'" said Chivers. There *is* a certain fluidity to female sexuality, at least. Since the 1950s, the standard method for measuring sexual orientation has been the Kinsey scale, a self-reporting system that ranks an individual from 0 (exclusively heterosexual) to 6 (exclusively homosexual), based on averaging out scores on four variables: attraction, fantasy, behavior, and self-identification. Most people—me included—score between 0 and 1 on the Kinsey scale: we are exclusively heterosexual, or nearly so.

But a survey of Australian twins revealed striking differences between men and women. While most of the subjects scored near 0, as expected, men scored at either extreme, either predominantly straight or predominantly gay. So the distribution curve for men peaked on one end of the Kinsey scale between 0 and 1, then dipped into a valley before peaking again at the gay end of the scale, between 5 and 6. However, the distribution curve for women showed a gradual tapering off as it moved from 0 to 6, with no peak at all at the gay end of the scale. So for women, there is more room for variation across a population. And a woman's orientation may shift over time, shaped by cultural influences, or positive or negative experiences.

Nixon's experience is not unusual. In *Living with Our Genes*, Dean Hamer related the tale of Margaret, a sixty-eight-year-old, twice-widowed woman who had led a happily heterosexual life. He thought she would score a solid 0 on the Kinsey scale, but as the session was ending, she told him he hadn't asked her about the future. "I may be sixty-eight, but I'm still interested in sex," she confessed, pronouncing most of the men within her age range "pathetic" as potential partners. "So I assume that my next lover will be a woman."

Margaret's sexual pragmatism reminded me of a scene I witnessed years ago in a Seattle bar, when an elegant young woman politely turned down a would-be paramour, only to have him snarl, "What? Are you, like, a lesbian?" She coolly looked him up and down, and drawled, "Are you the alternative?" While self-identified bisexual men do exist, they are rare, and it is difficult to imagine a straight sixty-eight-year-old man making a similar pronouncement, although one could argue about whether this is due to biological or cultural factors. The Australian data are based on self-reports, and while suggestive, they don't say much about the underlying genetic influences of this trait. But it seems that men are far more rigid than women when it comes to their sexual orientation.

Meredith Chivers has spent years studying sexual response in straight, lesbian, and bisexual women in the time-honored fashion of showing her subjects sexually explicit videos and measuring arousal. Not surprisingly, she found that women who identified as exclusively lesbian were more aroused by video imagery featuring women. She expected the opposite would be true for the women who identified as exclusively heterosexual: they would be aroused primarily by videos featuring men. Instead, straight women responded sexually to both male and female video imagery. They didn't differentiate at all. Sexual orientation in women does not appear to fit the male binary model; it is more of a continuum.

But these differences in sexual arousal are a matter of degree. "Gay men respond to opposite-sex stimulation too," said Chivers. It's just that there is a larger difference in the relative response to material that gay men prefer compared with something they don't prefer, whereas straight women "don't seem to differentiate in their physiological responses." This *might* be evidence of a fundamental difference between male and female sexuality, but that interpretation could change as cultural mores continue to shift. "There have been a lot of cultural taboos against identifying as bisexual among men," she said. "Maybe if we look at the data in fifty years, it will look very different."

Chivers also emphasized that just because something is happening physiologically within the brain, this need not be evidence of a "natural" state. "The reality is that our brain and physiological reactions are responsive to changes in our environment as well," she said. Women who feel aroused when viewing same-sex erotica are not necessarily bisexual, nor are those who score within the bisexual range on the Kinsey test or similar self-reporting instrument. They might identify as straight or lesbian or gender-queer. "Bisexuality is an identity level," said Chivers. "It's a way that an individual understands and formulates their sexuality. There is so much variability, especially among women, that goes into formulating that identity."

The public demanded Cynthia Nixon give herself a label, and she chose bisexual. But Nixon's broader point was a plea for accepting that the LGBT community is as diverse as the general population. Let us focus on people as individuals, she insisted, instead of imposing a label, or applying a litmus test for whether or not someone is truly gay. "Our community is not a monolith," she told the *Advocate*. And she is right. It is indicative of just how insufficient our binary model of sex and gender is, that even biological sex isn't as cut and dried as we usually assume.

Boy Meets Girl

The 1843 local election in Salisbury, Connecticut, might have been the closest in American history: it came down to a single vote, and a controversial one at that. The voter in question was twenty-three-year-old Levi Suydam, whose right to cast a ballot was challenged by the opposing party on the basis that he was "more female than male."* Suydam had narrow shoulders and broad hips—a decidedly effeminate build—liked bright colors and fabrics, and showed "an aversion for bodily labor." Yet when a physician named William James Barry examined Suydam, he found he had male genitalia. Barry declared Suydam male, and he was allowed to vote, tipping the election in his party's favor. A few days later, Barry discovered that Suydam's biological sex wasn't as straightforward as it seemed. Suydam had a vagina as well as a penis, and menstruated regularly.

Like Barry, most of us assume it is a straightforward matter

---

*Women were not granted the right to vote in the United States until 1920, when the Nineteenth Amendment was ratified.

to determine someone's biological sex. If someone has two X chromosomes, she is female; a person with one X and one Y chromosome is male. Most people fall squarely on the male or female ends of that continuum, but there are also so-called intersex people who are a little bit of both. Consider the case of Spanish Olympic hurdler Maria Martinez Patino, who should have competed in the 1985 Olympics in Kobe, Japan. Cheek swabs were administered to verify the biological sex of all athletes, as indicated by their chromosomes, and Martinez Patino's results came back XY instead of XX. Martinez Patino was disqualified—unfairly, many would contend. At issue is the definition of biological sex. She had a rare condition known as androgen insensitivity, in which the body lacks a key receptor to sense male sex hormones. Even though she had breasts and a vagina, she also had internal testes.

A similar case arose during the 2009 Track and Field World Championships, when South African sprinter Caster Semenya took the gold in the women's 800 meters and was promptly subjected to gender testing and forced to withdraw from competition. Like Patino, she is intersexed, with internal testes instead of ovaries. She produces more testosterone than most women, which helps build muscle. Some people think this gives Semenya an unfair advantage, although scientists from the Stanford Center for Biomedical Ethics disagreed, noting in a 2012 critique in the *American Journal of Bioethics* that testosterone is "just one element in a complex neuroendocrine feedback system" and is not the only factor in determining athletic performance. But the bias persists. "These kinds of people should not run with us," Italian sprinter Elisa Piccione griped to *Time*. "For me, [Semenya] is not a woman. She is a man."

The Olympic officials—and that spiteful competitor—

erroneously assumed that biological sex is wholly determined by chromosomes, but there are many other genes involved in development that contribute to whether a child is born male or female—or somewhere in between. Depending on the standard one uses, 4 percent of babies may be born intersexed, and often the condition is not detected until puberty. How does this happen? Remember that we inherit a set of chromosomes from each parent, including the XX or XY chromosomes that usually determine biological sex. But some people are missing one chromosome, or carry an extra one. And merely having the correct chromosomes isn't sufficient to determine biological sex, either.

There is almost no difference between male and female embryos upon conception. At eight weeks, a gene on the Y chromosome (if the fetus has one) called TDF switches on, producing a protein that activates another gene, which produces testosterone and other hormones to prevent female internal organs from forming. Hamer likens the process to a railway switchback: there are two tracks, male and female. If the TDF gene switches on, the fetus heads down the male track; if it doesn't, the fetus continues down the female track. But there are many other genes that must switch on at just the right time throughout the journey to ensure a fetus develops normally. For instance, a gene called r-spondin1 is linked to the development of ovaries. If that gene isn't functioning, that person, while genetically female, will grow up to be physically and psychologically male, although he will be sterile and it may not be clear whether he has male or female genitalia—or both.

Hence we find people like Emma, a patient of Johns Hopkins urologist Hugh H. Young in the 1930s, who possessed an abnormally large clitoris as well as a vagina, enabling her to engage in "normal" (penetrative) heterosexual sex with both men and women.* Emma was married, but admitted she didn't enjoy relations with her husband. She had many female lovers on the side, and despite her social status and outward appearance as a woman, her gender identity skewed male. But she declined Young's offer of corrective surgery to give her true male genitalia, declaring her vagina to be her "meal ticket." Her husband accepted her unusual condition and supported her well, and she didn't fancy having to work for a living. So was Emma a lesbian because she preferred sex with women? Was she a straight man who preferred women, but tolerated sex with her husband in exchange for domestic security? Was she bisexual? Is "she" even the correct pronoun to employ? Emma shatters our comfortable binary model by crossing all sorts of boundaries.

Historically, Emma would be classified as a hermaphrodite, a person with both male and female sex characteristics. Anne Fausto-Sterling, a prominent biologist specializing in gender studies and sexology at Brown University, famously proposed a tongue-in-cheek thought experiment in 1993 outlining a five-gender model. There are men, women, and "true hermaphrodites," defined as those with one testis and one ovary. There are also "pseudohermaphrodites": those whose external genitalia and secondary sex characteristics don't match their chromosomes. Male pseudohermaphrodites ("merms") are those with testes, XY chromosomes, and some female sexual characteristics, namely, a vagina and clitoris. They have no ovaries, nor do they menstruate. Female pseudohermaphrodites ("ferms") are those with ovaries, XX chromosomes, and some male sex-

---

* Meg Barker, among others, objects to the traditional definition of sex that limits it to strictly penetrative intercourse. In reality, human sexual behavior is very diverse.

ual characteristics—facial hair, a deep voice, and an adult-sized penis—but they are sterile and have no testes. Using this model, Patino and Semenya would be classified as "merms," while Suydam the ambiguously gendered voter may have been a "ferm." Even these additional categories don't encompass the many possible combinations. Fausto-Sterling acknowledged as much in a follow-up article, and has described biological sex as "a vast, infinitely malleable continuum that defies the constraints of even five categories."

Alas, we live in a binary world. The usual response to intersexed infants through much of the twentieth century was to perform surgery to "normalize" the genitalia, and raise them as one gender or the other. But this often backfires, as in the case of David Reimer. David was one of twin boys born in 1965. At eight months old, a minor medical problem made it necessary for the boys to be circumcised—except the doctor botched the job on poor David, burning off most of his penis. Johns Hopkins psychologist John Money recommended sex reassignment surgery for the boy, complete with surgical, hormonal, and psychological treatments. David became "Brenda." Money claimed Brenda as one of his greatest success stories. She "showed a clear preference for dresses over slacks," he wrote, liked her long hair, was neater than her brother, and didn't like to get dirty. While acknowledging that she needed to be trained to urinate from a seated position, as opposed to standing up, he claimed she liked to play with dolls and help her mother in the kitchen. What could be more "girly" than that?

Thirty years later, a follow-up study found that, far from liking dresses as a child, Brenda tore them off. She still tried to urinate standing up. She got into fights, preferred her brother's toys, and showed interest in a toy sewing machine only when she took it apart to see how it worked. When Brenda turned fourteen, she learned the truth about her birth, and opted to become David once again. He grew up to be a straight man, even marrying in 1990. Unfortunately, David Reimer's story doesn't end happily: he committed suicide in 2004.

Cases like his are one reason why Michael Bailey, a Northwestern University psychologist, and others, insist that sexual orientation and gender identity are determined by nonsocial factors. They contend that a large fraction of David's gender identity was already present at birth, so much so that an attempted sex reassignment, even as an eight-month-old infant, simply didn't take. Nature trumped nurture. "Innate traits emerge despite what happens in the social world," Bailey argued. "If you can't make a boy attracted to other males by cutting off his penis and rearing him as a girl, then social factors really can't be very important." William Reimer (no relation), a psychologist and urologist who has worked with more than one hundred cases of "sexual differentiation" disorders, reported that among all such cases—those with XY chromosomes raised as girls—the only ones who grew up to have female gender identities and to be attracted to men were those, like Patino and Semenya, who lacked the receptors for male sex hormones.

Yet it is never quite that simple. Gender identity, like sexual orientation, is subject to the same give-and-take between genes and multiple cultural and environmental factors throughout one's life. Children first become aware of gender between the ages of eighteen months and three years, and most of us develop a gender identity that matches our biological sex—but not always.

Mix and Match

In 2011, the UK's *Daily Mail* ran a front-page story about five-year-old Zach Avery. Biologically, Zach is a boy, but his brain insists he is a girl. His mother claimed he was quite "normal" until the age of four, when he became obsessed with *Dora the Explorer,* wanted to wear girls' clothing, and began insisting that he was a girl. Then the *Washington Post* weighed in with the story of a little girl named Katherine, who first started insisting she was a boy at the age of two. She wanted to wear pants and have short hair, and liked to play with trucks and swords. Her mother patiently tried to explain that she had girl parts and that made her a girl, but a confused Katherine responded, "When did you change me?"

Both Zach and Katherine exhibit symptoms of what the *Diagnostic and Statistical Manual of Mental Disorders* (DSM) calls gender identity disorder (GID): "a persistent and intense distress about the assigned sex, together with a desire to be (or insistence that one is) of the other sex."* Like gender nonconformity, this often manifests in atypical behavior, preferring the clothing and toys of the opposite sex and rejecting anything too boyish, in Zach's case, or girly, in Katherine's case. Merely exhibiting a few atypical behaviors is insufficient for a GID diagnosis. I was very much a tomboy growing up—early photographs show me with a pixie haircut, proudly brandishing a toy gun in a cowboy hat next to my older brother—but

---

*The language here is problematic. The literature routinely talks about such things in terms of "disease" or "disorder," when in fact it may be neither.

I had gender-typical interests too, and still self-identified as a girl. A child with GID, in contrast, exhibits "a profound disturbance of the normal gender identity"—at least according to the DSM.*

This raises some problematic issues for the parents of such children. Do they resort to therapy to try to stifle or change their child's behavior, thereby risking mental anguish? Or do they seek to accommodate the child's desire to be a different gender? Zach's parents and his school supported his gender-bending proclivities, although his mother noted that she made sure there were gender-neutral clothes in his closet should he decide he wanted to wear them. As for Katherine, she began attending school as Tyler. Both seem to have adjusted well to their new gender identities. The real dilemma will occur when the children reach puberty.

Ovid tells the tale of Iphis, in which a father swears he will only raise his unborn child if it is a boy; if his wife gives birth to a girl, he will kill the infant. When her daughter is born, the wife resolves to raise the child as a boy to save her life, on the advice of the goddess Isis. All goes well and the father is none the wiser, until Iphis reaches puberty and is betrothed to another girl. Desperate, the mother prays again to Isis, who transforms Iphis into a man. In the absence of divine intervention, however, parents and children must decide whether to take puberty blockers to stave off biological development, followed by hormone injections in their teens, anticipating possible sex-change surgery in adulthood.

This kind of preparation can make the transition easier, if that is what the grown child ultimately desires. However, be-

---

* The latest revision to the DSM employs more neutral language.

tween 43 percent and 80 percent of children with GID eventually grow out of it and do not retain their transgender identity into adulthood.* We are not born with our sense of self fully formed, after all. We arrive with certain traits in place that subsequently interact with our environment and experiences to shape the people we become. Most children and adolescents like to try on different roles before they settle into an adult identity. Fausto-Sterling told the *Washington Post* that gender identity may not even be fixed throughout life: "I doubt it is a permanent thing at age two." The hormone injections will make the child infertile, an irreversible side effect, and Fausto-Sterling has expressed concern over the unknown long-term consequences of these new treatments. So it is important to be absolutely sure that the transgender identity is permanent before following such a course of action.

Tyler's older sister blithely informed schoolmates that her "brother's" condition was "just a boy's brain in a girl's body." That is the customary framework applied when discussing transgender people, but this is once again foisting a binary model onto a reality that is much more of a continuum. Even within the transgender community, there is a great deal of diversity, because sexual orientation, gender identity, and gender expression are all separate dimensions. "Not all trans women are feminine, and not all trans men are masculine, and it's never as simple as someone being trapped in the wrong body of the assigned sex," writes Natalie Reed, a transgender woman who

---

*It can be difficult linguistically to talk about these issues. One source objected to my description of a child "reverting" back to his or her biological gender and suggested I replace it with "switched to the gender typically considered to be in line with their chromosomes or anatomy." More accurate? Yes, but far less clear.

blogs openly (albeit under a pseudonym) about her experiences, and cheekily refers to her genitalia as "Schrödinger's vagina."

Someone with male anatomy could be attracted to men (gay man), or could have a gender identity of woman (transsexual), or could occasionally indulge in female gender expression (cross-dresser). Someone with female anatomy could identify as a woman, have a masculine gender expression, and be attracted to women (butch lesbian). Emma, the hermaphrodite hausfrau, was biologically intersex, with a gender identity that skewed male, a feminine gender expression, and she was primarily attracted to women. As the Center for Gender Sanity states on its Web site, "It's a mix-and-match world and there are as many combinations as there are people who think about their gender."

Then there are those who are "gender-queer," a category of people who reject static notions of gender entirely, in favor of a more fluid blurring of the lines. Take the case of "Rory," who described himself as a "gender-queer person of the transgender female to male persuasion."* Rory was born female and grew up a rough-and-tumble tomboy in a South American country, immersed in what he described as a "very homophobic, transphobic, and gender-conforming society." He moved to the United States and became a scientist, but began seriously questioning his gender identity during graduate school. As someone who straddles both worlds, the binary division between male and female makes no sense to him. He is primarily attracted to other gender-queer people, followed by women and men, respectively. His gender identity skews male, and his gender expression skews female.

Rory is not a man trapped in a woman's body; he is someone

---

* "Rory" is a pseudonym to protect his privacy.

who embodies elements of both sexes and simply found it felt more "right" when people referred to him using male pronouns. He had his breasts surgically removed, because they felt "wrong," like they weren't really part of his body. In fact, Rory admitted that he spent much of his early life as "a walking brain," thoroughly dissociated from his physical body, unwilling to look at himself in a full-length mirror because the image it reflected didn't match the self-image in his own mind. This is a common experience for people in the transgender community, according to Natalie Reed, who writes that transgender people "are only seeking to get our bodies to conform to our sense of self, so that we can feel that they are our own, rather than a creepy gross alien thing that happens to be attached to us." She adds, "A body consistent with one's internal conception of sex and gender is a perfectly reasonable thing to want, and a very difficult thing to live without."

This disconnect between one's physical body and one's mental image of it might be related to our internal body maps. The brain generates these maps by integrating various sensory inputs, but sometimes there are gaps. There is a rare condition in which a person feels an overwhelming compulsion to amputate an otherwise healthy limb. The first recorded instance dates back to a 1785 medical textbook, describing the case of an Englishman who wished to amputate a leg after falling in love with a one-legged woman. The surgeon refused, despite the offer of one hundred guineas, and the man ended up forcing the surgeon to perform the operation at gunpoint. But it's not primarily a sexual fetish. Most patients insist that the body part in question has never truly felt like it was part of them in the first place, and they can identify down to the centimeter exactly where the limb "needs" to be cut.

Several years ago, neuroscientist Vilayanur Ramachandran conducted a simple experiment at the University of California–San Diego: he recruited patients who wanted to have a limb amputated, and recorded their brain activity while prodding the limb in question. He found that there was a response in that area of the cortex where sensory information is initially processed, but not in the region where such input is integrated to generate the body map. So there really is a discrepancy between the patients' mental body image and their physical form, and this naturally creates cognitive dissonance. Laura Case, a young researcher in Ramachandran's lab, has even uncovered preliminary evidence that many people who self-identify under the umbrella term "bigender," who also claim to switch involuntarily between male and female gender states, experience the sensation of having phantom breasts or genitalia, similar to the "phantom limb" phenomenon.

The Ramachandran lab has also reported lower rates of phantoms in transgender patients after surgery, and Case speculates that something similar could be at work in people like Rory, who never felt his breasts were part of his body. Perhaps that part of his anatomy never found its way onto his body map.

Rory said that he might start taking testosterone, just to see if it suits him. With his small stature, delicate frame, higher voice, and lack of facial hair, he is usually mistaken for a woman. "I suspect my gender identity will always be in that middle area," he acknowledged. "Without the effects of testosterone, I am read as a girl because society is really binary and puts you in these boxes."

Express Yourself

In 2011, the *Boston Globe* featured identical twin boys, Patrick and Thomas, who share the same genes and home environment, yet could not be more different in terms of their behavior. By age three, Thomas liked toy guns, while Patrick liked Barbie dolls and wearing his mother's shoes. At five, Thomas wanted to be a monster for Halloween and Patrick wanted to be a princess, although he switched to Batman after being told other kids would laugh at him. Patrick always preferred playing with girls instead of boys. Such a boy is far more likely to be diagnosed with Childhood Gender Nonconformity (CGN)— when a child's gender expression doesn't mesh with the typical behavior for that child's biological sex. According to Bailey, there is a very good chance Patrick will grow up to be gay; the data suggest that 75 percent of boys diagnosed with CGN turn out to be gay or bisexual, and Bailey thinks that percentage might be even higher.

There is a problem with this working premise: who decides what is "masculine" or "feminine" behavior? One of Michael Bailey's more questionable assertions in his book *The Man Who Would Be Queen* is that dance—ballet in particular—is an innately "feminine" activity. It is true that roughly 50 percent of male ballet dancers are gay. Yet I find Bailey's assertion problematic because there is no acknowledgment of potential cultural influences: there may be a higher percentage of gay male ballet dancers because such men are welcomed in that profession. Neither does this stance take into account the 50 percent of male ballet dancers who are straight, or the many cultures around the world where dancing is a perfectly appropriate activity for men.

The truth might lie somewhere in between. My straight brother has always been ruggedly masculine, playing with the expected toy trucks and guns and helping my father build things around the house. But he also loved to help my mother cook in the kitchen. When my mother tried to teach me to knit and crochet, I failed miserably, producing a gray misshapen mass of yarn that bore no resemblance to the intended scarf, but my brother took to it like it was the most natural thing in the world. His interest had nothing to do with a masculine or feminine label for a given activity. He just liked making things with his hands. Today he builds houses, drives a semi, and also bakes pies from scratch for family holiday dinners. I'm sure he could still crochet an Afghan throw in a pinch.

The situation is more muddled when it comes to gender nonconformity in girls. A woman's sexual orientation may be partly influenced by prenatal exposure to a male sex hormone called androgen, and women exposed to greater levels of the hormone in the womb may exhibit more gender nonconformity in childhood. But this behavior does not necessarily correlate with a woman's sexual orientation later in life. Granted, some tomboys grow up to be lesbians, or bisexual, or even transgender men, but most grow up to be straight women who just happen to gravitate toward non-girly things. "Having a tomboyish girlhood is not going to be predictive of sexual orientation," said Meredith Chivers.

Despite a certain fluidity in pronouns, there is no indication in Heldris of Cornwall's text that Silence is confused about her gender identity or her sexual orientation: she rejects the queen and marries the king, after all. Fundamentally, her story is about the struggle against the constraints of stereotypical gender roles: the activities, expectations, and behaviors that any given society assigns to its men and women. She simply wants to be able to hunt, ride, and fight as she always has, without having to pretend that she is something she is not. And me? I never wanted to be a boy, and neither did I secretly harbor same-sex desires. I just found the socially acceptable options available to me as a girl to be too limiting. I wanted to have swashbuckling adventures and save the damsel in distress, not wait around to be rescued myself. There is a saying that nine-tenths of the law of chivalry is the desire to have all the fun, and any inveterate tomboy would agree with that.

"I absolutely think that when you're talking about masculine or feminine, you're talking about a continuum, but there's nothing more humbling than having a child," said Chivers, who admitted that her own son challenged her assumptions. Despite the fact that she and her partner provided a gender-neutral household and never taught him anything about shooting guns or fighting, he loves those forms of play. That said, "Culture is the arena in which we express all of this, and there are many variations, and expectations, of gendered behavior," she said. "There is so much societal pressure to conform to those proscribed gender roles."

Deeply ingrained attitudes about gender infuse every aspect of our society, including the marketing of children's toys. In 2012, a New Jersey eighth-grader named McKenna Pope made headlines when she lobbied the toy manufacturer Hasbro for a gender-neutral version of its classic Easy-Bake Oven for her four-year-old brother, who aspired to be a chef. The toy is decorated in pink hearts and flowers and sold on Hasbro's Web site under "Cooking and Baking Games for Girls." But there is nothing intrinsically feminine about cooking—just ask the top chefs in the world, the majority of whom are men.*

Do the clothes make the man, or vice versa? Legend has it that punk rocker Iggy Pop's manager once showed up to bail his client out of jail after Pop was arrested for drunk and disorderly conduct, and asked why he was wearing a woman's dress. "I beg to differ, this is a man's dress," Pop replied. Actor and comedian Eddie Izzard has notoriously dressed in women's clothing on-stage, insisting, "They're not women's clothes, they're my clothes. I bought them." Both Pop and Izzard are straight, although Izzard has described himself as a "straight transvestite or male lesbian" or "complete boy and half girl." An eight-year-old "pink boy" named P. J. proved wise beyond his years when he told the *New York Times* about a boy in his class who loved soccer: "He comes to school every day in a soccer jersey and sweat pants, but that doesn't make him a professional soccer player."

It is conventional wisdom that one has either a male or female brain, and perceived differences between men and women in personality, cognition, emotion, and behavior are often cited as evidence of this. I confess to being rather sensitive to this issue. Nothing raises my hackles more quickly than an overconfident pronouncement about how men are just naturally better than women at X, or how women have a more natural affinity for Y—because gender stereotypes swing both ways. I could easily point to several exceptions to any given pronouncement, because people are individuals, not averages. On average, men are taller than women. But there are plenty of shorter men and

---

* It has been said, only half in jest, that when women cook, it's a domestic chore; when men cook, it's professional haute cuisine.

taller women, so it would be foolish to use height to make a prediction about any one person's biological sex. Why should cognitive ability, aggression, play behavior, or empathy—to name a few perennially cited traits—be any different?

Yet it would be equally foolish to deny that there are biological differences between men and women—the hormones alone see to that—and these differences are especially pronounced when it comes to anything related to sexual and reproductive behavior. Case in point: researchers at the University of California–San Francisco identified several genes linked to testosterone and estrogen and found that these control gender-specific behavior in mice, notably their sex drive, aggression, and parenting style. Female mice unable to sense the hormone estrogen lost interest in sex and in caring for offspring. Male mice fortified with testosterone were more aggressive toward other males, tried to mount female mice, and marked their territory with urine. When testosterone was taken away, the castrated male mice did not behave as aggressively.

But here is the interesting bit: The researchers could also knock out some behaviors and preserve others, such that male mice showed altered mating behavior but still wanted to fight and mark territory. Female mice still showed interest in sex but spent less time caring for their young. UCSF's Nirao Shah, who led the study, suggested that perhaps a sex hormone is like the main breaker that regulates electricity to the entire house. Individual genes influenced by sex hormones are like the light switches in each room, making it possible to turn the lights on in the kitchen while leaving the bedroom dark.

Brain-imaging studies do show differences in male and female brain structures: size, specific brain regions, neurotransmitter content, number of receptors, composition of neurons, and so forth. We also know that hormone levels can vary significantly among individuals. But we know very little about how those differences translate into behavior. True, these differences appear to play a role in sexual orientation and gender identity, but the further you go from reproductive behavior, the less significant the sex differences, according to Donald Pfaff, a neurobiologist at Rockefeller University. Play behavior differs about the same amount as height: boys on average will choose a truck, compared to girls, who prefer dolls, although there are enough exceptions to that rule that is it not possible to predict sex from a child's play preferences alone. Other qualities such as empathy, or spatial skills, show roughly half the heritability of height. Once you control for social and cultural factors, there is almost no significant variation between the sexes when it comes to verbal and math ability.

Perhaps it's not so much a binary case of a male versus a female brain, as it is another continuum. In 2001, psychologist Daphna Joel of Tel-Aviv University made a provocative argument for brains being intersex, drawing on a plethora of studies to bolster her case. She speculated that sex interacts with other factors, both in the womb and throughout life, and the structure of the brain is thus constantly reshaping. The result, Joel argued, is that "human brains are composed of an ever-changing heterogeneous mosaic of 'male' and 'female' brain characteristics, rather than being all 'male' or all 'female.'" In short, different individuals will have different combinations of male and female characteristics.

Maybe we all need to follow Cynthia Nixon's example and push back more strenuously against the binary bias that pervades our culture when it comes to sexuality and gender. Such stereotypes arise from lazy thinking, and while they might make it easier to deal with the complexity in the world, they also make it far too easy to lose sight of people as individuals.

Nobody knows this better than graphic artist and photographer Mark Rodriguez. His twin daughters shared a room but had very different tastes: one liked princesses and the other was obsessed with superheroes. Rodriguez decorated each half of the room according to each girl's taste, photographed it, and posted the images online, whereby someone commented, "It feels like the girl's side of the room takes up just a little more than the boy's side." To which Rodriguez politely replied, "Actually, they are both 'girl' sides. I have twin daughters and one is infatuated with Superman."

III

WHY

7

Feed Your Head

Mrs. Coplin: Why are you not wearing pants?
Agent Paul: I had an experience, that's why. I re-
sisted at first, and then it evolved, and it continues to
evolve for me.

—*Flirting with Disaster* (1996)

placed my hand on the coffee table, fingers outspread, and
watched in fascination as blue veins seemed to rise and float
above the surface while an intricate web of colorful geometric
shapes swirled across my skin like a dynamic tattoo. Then the
edges dissolved into the table's wooden surface, and those edges
in turn melted into walls that seemed to ripple and pulse. When
I closed my eyes, a vast, swirling kaleidoscopic scaffolding of
geometric patterns snaked across my lids. I was Alice, tumbling
down the rabbit hole into Wonderland, a waking dream state
where there were no hard edges separating self and object, no
cohesive sense of my own body in space and time, as if my molecules had been dispersed into the universe to mingle freely with
the molecules of everything else. I could, with great effort, open
my eyes and will my molecules back into some loosely bonded

semblance of "me," but the real world seemed harsh and over-bright. It was so much easier to close my eyes and sink back into the waiting embrace of the psychedelic void.

My mad excursion into Wonderland started just a few hours earlier, with a little white tablet that looked and tasted like a peppermint Altoid. But no mere breath mint ever broke down the walls of anyone's reality so completely. Historically it has gone by many nicknames: purple haze, electric Kool-Aid, Looney Tunes, windowpane, blue cheer, and white lightning. These days it's often called doses, trips, Sid, or Orange Sunshine. All refer to the same substance: lysergic acid diethylamide (LSD), or acid, a member of the tryptamine family that is derived from ergot, a grain fungus commonly found in rye.

LSD belongs to a class of drugs known as psychedelics, coined from the Greek root words for "mind revealing" because of their ability to alter cognition and perception. Its best-known siblings include psilocybin, the active ingredient in magic mushrooms; 3,4-methylenedioxy-N-methamphetamine (MDMA), better known as Ecstasy; peyote, made from the ground-up tops of a cactus called *Lophophora williamsii* that contain mescaline; and ayahuasca, a bitter tea made from the Brazilian vine *Banisteriopsis caapi* (dubbed the "spirit vine"), whose primary active ingredient is dimethyltryptamine (DMT). Most are classified as Schedule I substances by the U.S. Drug Enforcement Agency and are illegal, but that hasn't dampened their popularity.

Ecstasy is by far the most popular, but LSD has done a brisk business over the years, perhaps because it is fairly easy to make. Initially, people soaked sugarcubes in the substance, and then started encapsulating LSD in pill form, like my little mint. Next came thin squares of gelatin, dubbed "windowpanes," before

makers hit on the idea of using blotter paper: sheets of paper soaked in LSD, which is dried and cut into small individually dosed squares that the user could simply pop onto his or her tongue. This approach appealed artistically, as well, since the various distributors could print their own unique designs on the blotter paper as a kind of trademark. It is an art form all its own.

LSD is water soluble, odorless, colorless, and tasteless, and a dose as small as a single grain of salt (about 10 micrograms) can produce mild effects, with the full mind-altering impact kicking in at higher doses (between 50 to 100 micrograms). A little bit of acid goes a long way. A chemist named Augustus Oswald Stanley—immortalized in Tom Wolfe's *The Electric Kool-Aid Acid Test*—was arrested in 1967 with sufficient LSD in his possession to concoct a quarter of a million acid trips, and claimed with a straight face that it was for "personal use." The judge must have bought this argument, because his sentence was just two years. Stanley would not likely get off so lightly today. Because it is chemically unstable and degrades quickly, LSD is usually contained within a medium—a neutral liquid or blotter paper—and sentencing is based on the total mass of the drug plus its medium, even though one might technically possess only a small amount of pure LSD.

The psychedelic effects last several hours and include vivid Technicolor hallucinations, warped perception of time and space, and what is often described as a temporary dismantling of the ego or sense of self. When the drug finally captured widespread public attention, it transformed popular culture, defining an entire era. Artists on acid gleaned inspiration from their inner visions, producing the famous psychedelic imagery on concert posters and album covers for rock bands in the 1960s and 1970s. As for the musicians—well, I suspect the four-minute keyboard solo in the middle of the Doors' "Light My Fire" is evidence that the band members were tripping and lost track of time. Iron Butterfly's "Inna Gadda Da Vida" runs a whopping seventeen minutes, with a lengthy drum solo to boot. Both tunes make me want to pound my head against the wall to make the cacophony stop.

Perhaps my failure to appreciate Iron Butterfly stems from the fact that I don't do drugs. This has nothing to do with any moral qualms; I just prefer my reality straight up, no chaser. Also, I don't like inhaling smoke into my lungs, snorting powder up my nose, or sticking needles in my body for anything other than necessary medical procedures, which rules out most of the usual suspects. My one pathetic attempt at smoking pot in college ended with me refusing to inhale, thereby earning the ridicule of my peers, plus it took a week to get the smell out of my clothing. Munching on pot-laced brownies made me sluggish, sleepy, and a bit queasy, with no hint of the promised "high." Frankly, the whole experience bored me. As with alcohol, my body seems to be wired in such a way that any buzz I might get from the forbidden substance du jour is simply not sufficient to compensate for the inevitable physical unpleasantness of the aftermath.

But acid seemed like it might be different. The delivery mechanism is refreshingly simple: just pop it in your mouth and let it dissolve—no snorting, inhaling, or injecting—and wait for the psychedelics to kick in. I'd been intrigued by the stories of its mind-expanding properties from various acolytes over the years, but they were so irritatingly smug and pushy about making converts, my natural response was to dig in my heels and resist. Then I encountered a charming elderly woman who lived upstairs from me in Washington, D.C. We sat on her pink floral

sofa and she served me Earl Grey tea in dainty cups resting on lace doilies as she blithely told me how she left her husband and ran away to Berkeley in the 1970s to "turn on, tune in, and drop out," amassing some surprisingly salacious adventures along the way. Acid, she claimed with great sincerity, had changed her perspective on life.

The clincher came one night over dinner with friends, when I mentioned I was exploring the science of self, and the many different ways we construct identity. "In that case, you should really try acid," one man advised. He was a child of the Wood-stock generation and had just related a hilarious tale of getting the neighbor's cat stoned on pot by placing the cat inside a paper bag and blowing smoke into it. Apparently felines have an affin-ity for weed: the next day he found the cat sitting patiently in-side the paper bag, jonesing for the next hit.

He described how acid dismantles the ego, and opens a win-dow into a hidden reality lurking just beyond human percep-tion. While he insisted that LSD is not for everyone, he felt confident I could handle it.

"Handle what, exactly?" I asked.

Some people found the warping of reality and melting of boundaries extremely upsetting, he explained, adding that once the drug wore off, "The self comes roaring back with a ven-geance. And believe me, you'd better like what you see."

That did it. I had to experience this firsthand. After all, it was "research." A few months later, I found myself at a secluded beach house for the weekend with my husband, Sean, on hand for moral support. My self-appointed guru of psychedelics had thoughtfully laid out three tabs of acid, two tiny blue Valium—in case I found myself on a bad trip and needed to come down quickly—and one final piece of advice: "Small things have a way of disappearing when you're on acid. Like keys. So don't worry about locking up."

Within thirty minutes, I started to feel queasy and light-headed. "I hope this isn't going to be just like the flu," I dutifully wrote on my notepad, followed by "My teeth feel funny." Restless, we went for a walk on the beach. It took forever, and yet no time at all. I found that if I stopped and focused intently on the sand, it started to ripple, as if there were snakes slithering just beneath the surface. "I think we need to get back to the house," I muttered, and we took refuge in the walled-in garden, where a large boulder pulsed as if it were breathing, and the roar of the surf pounded in my ears. Sean dutifully recorded the most boring video footage ever: me staring fixedly at a single wooden slat in the deck chair for several minutes, motionless, before looking up and announcing to the camera, "You have to go *into* the wood, down to the molecules." When we went inside, I promptly laid down on the Oriental rug and let myself be absorbed into its intricate patterns.

It proved impossible to take many notes. My hand kept melding with the pen and paper. These were the last words I managed to scrawl before surrendering to sensation and letting reality slip away: "Molecules. I keep going on about molecules but that's what it's like. Boundaries fade the longer you stay in one place. Every object is a living, breathing thing and we are all part of it. Oh god, I sound like every stoner acid head ever."

Doors of Perception

One fine April day in 1943, a Swiss chemist named Albert Hoffmann was tinkering in his lab, recrystallizing an unusual chemical compound derived from ergot that he had first synthesized

five years earlier—lysergic acid (the molecular core of most ergot alkaloids) combined with a diethylamine building block that he had tagged LSD-25. It hadn't seemed very interesting at the time to his employer, Sandoz, a giant pharmaceutical company interested in exploiting the drug potential of ergot-derived compounds. But Hoffmann decided to revisit LSD-25 on a hunch, struck by the similarity of its chemical composition to a well-known drug to increase blood flow.

In the Middle Ages, ergot-contaminated rye bread caused numerous outbreaks of a disease called St. Anthony's Fire, which took one of two forms. Symptoms of gangrenous ergotism included a burning of the skin, blisters, and dry rot in fingers and toes. The first known mention of this form dates back to 857, in a German historical treatise describing an outbreak near Cologne: "a Great plague of swollen blisters consumed the people by a loathsome rot, so that their limbs were loosened and fell off before death." Convulsive ergotism, meanwhile, primarily affects the central nervous system, causing mania, psychosis, convulsions, hallucinations, and prickling sensations. Some scholars have speculated that this form of ergotism was the underlying cause of accusations of bewitchment during the Salem witch trials, although the hypothesis is disputed.

Yet there is a kinder, gentler side to ergot: for centuries, midwives used it as a natural remedy to control bleeding after childbirth. Indeed, ergot-derived medicines for postpartum bleeding and migraine headaches comprised one of Sandoz's most lucrative markets in the 1940s.

Working with ergot was tricky business. In its basic alkaloid form, it proved complex and chemically unstable; a single mistake could cause it to break down into black tar. It was also potent stuff: Hoffmann was handling a mere few hundredths of a gram, yet as he worked, he became restless and dizzy, so much so that he gave up and went home to bed. There, he experienced a fitful, "dreamlike state" filled with intense colors and "extraordinary shapes." Somehow, he had exposed himself to trace amounts of LSD-25, likely absorbed through the skin, and even that trace was sufficient to induce mild hallucinations. A curious Hoffmann repeated the experiment three days later, with a much larger dose of 150 micrograms.

That day—April 19, 1943—became known in his journals as Bicycle Day, because Hoffmann was so incapacitated by the dosage that he had to ask for a ride home on his lab assistant's bicycle. The journal entry for that day is sparse—some fifty words describing the dosage and the peculiar sensations he experienced ("dizziness, feeling of anxiety, visual distortions, desire to laugh")—since Hoffmann struggled as much as I had to write once the drug kicked in. Initially, he had a pretty bad trip: he later described feeling as if his disembodied ego was suspended in time and space, and becoming convinced that his kindly next-door neighbor was "a malevolent insidious witch with a colored mask." When his assistant brought in a doctor to examine him, the physician found no obvious signs of illness, other than dilated pupils and a slightly weakened pulse.

After a few hours, the horror subsided, replaced by the same geometric light show whenever he closed his eyes that I experienced. "Kaleidoscopic, fantastic images surged in on me, alternating, variegated, opening and then closing themselves in circles and spirals, exploding in colored fountains, rearranging and hybridizing themselves in constant flux," Hoffmann wrote in his memoir. There was even a touch of synesthesia: "Every sound generated a vividly changing image with its own consistent form and color."

Hoffmann described his acid experience as viewing reality via a different wavelength. Just as there are many different wavelengths in the electromagnetic spectrum, he mused, so too there are many different realities, "each comprising a different consciousness of the ego." This, Hoffmann felt, was the true importance of LSD: it biochemically alters the brain's "receiver" and tunes it to a different wavelength.

"LSD is a chemical lens that allows you to see things in a different dimension," agreed University of South Florida clinical neurologist Juan "Zeno" Sanchez-Ramos, who readily cops to having dropped acid as a young expatriate in Paris in the 1960s and credits those experiences with igniting his interest in the intricate workings of the brain. "You see the relationship between mind and matter, body and soul." He compared the drug to a microscope, opening up a new world otherwise invisible to the naked eye. Telescopes tuned to different wavelengths of light—radio waves, X-rays, gamma rays, and infrared—revolutionized astronomy by revealing all kinds of useful information about the universe we would not be able to perceive using visible light. Psychedelics might revolutionize neuroscience, if they can just shake off that '60s stigma.

This is why the psychedelic artists of the 1960s drew so much inspiration from drug-induced hallucinations—and why acid trips continue to inspire artists today. In 1989 a psychiatrist in Los Angeles named Oscar Janiger asked a group of artists to create paintings depicting their choice of subject, and then to re-create the same painting while on acid. Nearly all of them felt the LSD enhanced their creativity.

This perceived benefit isn't limited to artists. The late Steve Jobs once claimed, "Doing LSD was one of the two or three most important things I have done in my life." There is a reason

Apple's mantra is "Think different." (Jobs's Microsoft rival, Bill Gates, also dabbled with LSD, but dismissed it as youthful exploration.) Several notable scientists experimented with the drug, including Francis Crick, who once claimed he had "perceived the double-helix shape [of DNA]" while tripping.* John C. Lilly, inventor of the first sensory deprivation chamber, and Kary Mullis, whose improvements to the polymerase chain reaction (PCR) technique revolutionized DNA sequencing, also admitted to dropping acid. Physicist Richard Feynman was initially reluctant, worried it might damage his big mathy brain, but curiosity won out in the end: he, too, visited Wonderland.

Counterintuitively, this boost to human perception might be due to a decrease in brain activity, rather than an increase. In *Doors of Perception*, one of the most influential texts of the psychedelic movement, Aldous Huxley opined that LSD works by counteracting a "reducing valve" in the brain that normally limits our perceptions, such that we get far more sensory input than usual when on acid. Normally, we experience "a measly trickle of the kind of consciousness" necessary to "help us stay alive." His words seem eerily prescient in light of the surprising conclusions of a 2012 British neuroimaging study.

Neuroscientist Robin Carhart-Harris and his Imperial College colleague, David Nutt, used fMRI to scan the brains of thirty volunteers—all experienced users of psychedelics—while under the influence of psilocybin. They found that overall brain activity dropped, compared to scans taken after ingesting a saltwater placebo, especially in certain heavily interconnected

---

* Crick did not explicitly claim that acid had inspired the actual 1953 discovery of the DNA double helix, merely that being in that altered state allowed him to "see" it in his mind's eye.

"hubs" in the sensory regions of the brain that help constrain the way we see, hear, and experience the world, grounding us in reality. Those hubs "act like traffic circles that link disparate regions," Caltech neuroscientist Christof Koch explained in *Discover*. "The brain on psilocybin becomes more disconnected, more fragmented, which might explain some of the dissociative aspects of acid trips. Yet why this state should cause the mind-expanding effects is utterly unclear."

Neuroscientists refer to this collectively as the default network, and there is ample evidence that it plays a critical role in our sense of self. "We only realize we have an ego when we lose it," said Carhart-Harris. "That is the shocking and revelatory thing that people experience with psychedelics. We have this unconscious feeling that we have always been who we are, and we don't realize that we are a construct. So it can feel as if there is confusion about where we end and the world begins, as if we kind of merge with the world"—or, in my case, with the Oriental rug.

Eyes Wide Shut

Speculation about surrealist artist Salvador Dalí's possible use of LSD abounds, given the melting clocks, vibrant colors, and warped landscapes found in his canvases, but LSD wasn't discovered until the late 1930s and didn't become popular as a recreational drug until the 1960s. Dalí's artwork was weird long before that. The artist liked to experiment, but he didn't advocate habitual drug use: "Everyone should eat hashish, but only once," he purportedly observed, insisting he didn't need drugs to be creative. Timothy Leary, the groovy guru of psychedelics, admitted that Dalí was the only person who could paint LSD without having to take LSD. I couldn't find any record that Dalí ever tried acid, but I wouldn't be surprised if he had—just once.

There was definitely a touch of Dalí in my own experience as once familiar objects took on all kinds of reality-bending elements. As with an optical illusion, all I had to do was look at something intently and let my eyes unfocus a bit. A painting in the living room gained depth and motion, while the crystals in the kitchen's granite countertop swirled in pretty spirals. Yet as much as I marveled at these effects, it made my head ache to keep my eyes open for too long. Closing them made the ache go away, and intensified the hallucinations, creating a vivid interplay of complex colors and patterns. "You have to close your eyes to really see," I confided to a mystified Sean, burying my face in his shirtfront. It's possible I drooled a little. "Your eyes just get in the way." At which point he morphed into a green-and-purple Dragon-Man.

The idea is less drug-addled than it sounds. There is something about the sensory input we process via the visual cortex that is central to our sense of self, anchoring us in the real world. Vilayanur Ramachandran has opined that the line between perception and hallucination is blurrier than we think. "Both hallucinations and real perceptions emerge from the same set of processes," Ramachandran wrote in *The Tell-Tale Brain*. "The crucial difference is that when we are perceiving, the stability of external objects and events helps anchor them. When we hallucinate, as when we dream or float in a sensory deprivation tank, objects and events wander off in any direction."

We rarely stop to think about what an incredibly complex process happens inside the brain every time we observe the world. Light reflects off the objects around us, enters through our eyes, and is focused onto the retina. The retina is lined with photoreceptor cells that convert the light into electrochemical signals. These travel to the brain and stimulate neurons in the visual cortex in patterns that, under normal circumstances, mimic the patterns of light reflecting off objects in your field of vision. So the patterns of excited neurons ultimately form the images that you "see." Numerous brain-imaging studies have shown that psychedelics disrupt that normal brain activity and boost the random firing of neurons in the visual cortex, so your brain is getting both external and internal input signals.

Back in the 1920s and 1930s, a University of Chicago neurologist named Heinrich Klüver classified hallucination patterns into tidy categories known as form constants: checkerboards, honeycombs, tunnels, spirals, and cobwebs. More than seventy years later, another Chicago researcher, Jack Cowan—who holds dual appointments in mathematics and neurology—and University of Pittsburgh mathematician and computational biologist Bard Ermentrout set out to reproduce those hallucinatory patterns mathematically, believing they could provide clues to the brain's circuitry. While the random fluctuations in brain activity might technically just be "noise," the brain will take that noise and turn it into a pattern. Since there is no external input when the eyes are closed, that pattern should reflect the architecture of the brain, specifically the functional organization of the visual cortex. Robin Carhart-Harris described that organization as being fractal in nature, repeating the same patterns at different size scales. "Like tree branches, the brain recapitulates," he said. "You are not seeing the cells themselves, but the way they're organized—as if the brain is revealing itself to itself."

Cowan found that the predicted patterns from his calculations closely matched what people see when under the influence of LSD, and suspected these patterns might arise from the so-called Turing mechanism, named in honor of Alan Turing. Turing is best known scientifically for his work on codebreaking, computing, and artificial intelligence during World War II, and personally for his arrest, conviction, and chemical castration by the British government because he was a homosexual, leading to his suicide in 1954. In the midst of all that personal drama, however, he still found time to publish a seminal paper in 1952 on the mathematics of certain regular repeating patterns in nature, notably tiger stripes, leopard spots, and the precise spacing in rows of alligator teeth. These are known as Turing patterns.

Turing came up with a set of equations to account for such patterns. He proposed that the patterns arise from interactions between two chemicals that spread throughout a system much like gas atoms in a box do, with one crucial difference. Instead of diffusing evenly like a gas, the chemicals diffuse at different rates. One chemical acts as an "activator," while the other acts as an "inhibitor." The activator chemical expresses a unique characteristic, like a tiger's stripe, and then the inhibitor chemical kicks in periodically to shut down the activator.

Mathematical equations are very flexible. The same type of mechanism might determine the distribution of species in certain ecological systems, most notably the predator-prey model, in which the prey function as activators, seeking to reproduce and increase their numbers, while the predators act as inhibitors, keeping the population in check. Neurons, too, can be described mathematically as activators or inhibitors, encouraging or dampening the firing of nearby neurons.

Neurons respond not just to color and brightness in the visual field—the external input—but also to internal interactions with other neurons. Nigel Goldenfeld, a physicist at the University of Illinois, Urbana-Champaign, worked with Jack Cowan on a model for a generic neural network with random connections. They showed that in such a case, the firing of neurons would amplify the Turing effect, making hallucinations more common. But if our visual cortex actually behaved in this way, it would interfere with our vision. "You don't want to be enthralled by a hallucinatory spiral when there is a dangerous tiger in front of you," said Goldenfeld. So he and Cowan speculated that this might be why our brainy architecture is non-random: it confers an evolutionary advantage that limits interactions to stronger short-range connections with nearby neurons. Excited neurons simply follow the familiar uniform diffusion patterns we associate with the behavior of atoms in a gas, and the visual external input from the eyes easily dominates any weaker internal activity.

It could be that LSD is literally mind-expanding, in that it extends the interactions between neurons across longer distances, thereby producing the geometric patterns I saw whenever I closed my eyes. Once the drug wears off, those long-distance interactions no longer occur. As loopy as it sounds, acid really could give you a new way of "seeing"—and what you are seeing may be the synaptic architecture of the brain itself.

Jagged Little Pill

It never occurred to Albert Hoffmann that LSD-25 would ever be used as a recreational drug. He was primarily interested in potential medical applications. The Sandoz pharmacologists who initially dismissed the compound weren't looking for the right chemical properties. Its true potential lies in how it works in the brain—specifically how it affects serotonin, which was discovered around the same time as Hoffmann's psychedelic adventure.

The brain controls perception and communication throughout the body via chemical neurotransmitters that relay vital information from neurons. Each neurotransmitter attaches to matching areas on nerve cells known as receptors, thereby changing the behavior of the cell in some way. Typically, serotonin molecules pair with serotonin receptors, but LSD disrupts this tightly controlled communication system because its core chemical structure is so close to that of serotonin. Acid can attach to many of the same receptors, altering how someone perceives the world.

Sandoz launched the drug for the treatment of psychiatric disorders in 1947 under the trade name Delysid, because of its similarity to serotonin, which is linked to anxiety and depression. Between 1950 and 1965, more than 1,000 research papers were published, and some 40,000 people around the world were treated with LSD. A California radiologist named Mortimer Hartman left his thriving practice after experiencing acid and teamed up with psychiatrist Arthur Chandler to set up the Psychiatric Institute of Beverly Hills. They charged $100 a pop for acid-augmented sessions, earning the scorn of Aldous Huxley. "Really, I have seldom met people of lower sensitivity, more vulgar mind," he sniffed.

The institute was a smashing success: at its peak, the treatment rooms were fully booked five days a week. Hollywood luminaries such as Cary Grant, Betsy Drake (Grant's wife at the time), Esther Williams, and Judy Balaban (daughter of Paramount Pictures president Barney Balaban) began taking LSD as part of their psychotherapy. Most reported positive experiences and found the drug enhanced their progress. "All my life I've been going around in a fog," Grant purportedly told Williams. "You're just a bunch of molecules until you know who you are." But he also sounded a cautionary note: "It takes a lot of courage to take this drug. It's a tremendous jolt to your mind, to your ego."

Setting is a critical factor, since LSD and similar drugs can elicit very different experiences under different circumstances. Each room at the institute featured a couch in the corner. Subjects would lie down, put on a blindfold to block out distractions, and then take the prescribed pills and bliss out for several hours while soothing music played in the background. Nevertheless, the occasional patient still experienced a bad trip. Actress Marion Marshall initially panicked when her hallucinations revealed an enormous black-widow spider poised to attack, but rather than ending the session, she opted to face down her fears. She later declared it the best session she ever had. Unfortunately, Chandler became a bit too fond of the drug himself and was often tripping when he should have been looking out for his patients. It didn't inspire confidence in actress Polly Bergen, for one, who found another therapist.

Not all of the early research was so well intentioned. The CIA notoriously experimented with LSD as a possible mind-control drug during the Cold War, establishing a project called MKUltra, led by a chemist with a metaphysical streak named Sidney Gottlieb. The drugs were given to unwitting subjects, often prostitutes and their clients, mental patients, and convicted criminals—people who, as one of Gottlieb's colleagues observed, "could not fight back." A habitual user himself, Gottlieb set up a string of CIA-controlled brothels in San Francisco, in which prostitutes would slip drugs to their customers and agency officials would observe the results through two-way mirrors. One mental patient received doses of LSD for 174 straight days, and at least one subject died. The project was an unqualified failure. LSD, Gottlieb found, might make people more suggestible under its influence, but it couldn't be used for mind control.

All this fed into growing fears about the unpredictability and safety of psychedelics, which gradually approached hysteria. Rumors abounded of horrific trips, states of temporary psychosis, and flashbacks in which past acid-induced hallucinations recurred months or even years later. Concerns arose that LSD might cause genetic damage, particularly in unborn children, and the *Journal of the American Medical Association* ominously warned that psychedelic drugs could cause permanent "personality deterioration." LSD also became associated with the Beatnik movement of the 1950s, whose acolytes experimented with all sorts of substances.

The brewing culture war was further stoked by the antics of Timothy Leary. Originally a psychology professor at Harvard, Leary began conducting experiments with psilocybin in 1960 after having tried magic mushrooms on a vacation in Mexico. His initial research results appeared promising; the vast majority of his test subjects said they would repeat the experience. But other faculty members expressed serious reservations about Leary's research. Then he discovered LSD. Harvard students began making like Beatniks, taking magic mushrooms and LSD recreationally, outside the bounds of the official study. Alarmed parents and influential donors registered complaints. Tensions escalated until 1963, when Leary and his collaborator, Richard

Alpert, were dismissed from Harvard, ostensibly (in Leary's case) for not fulfilling his class lecture obligations.

With the aid of a wealthy benefactor, they set up a private research program in a New York mansion called Millbrook, although the "experiments" lacked protocols and amounted to little more than LSD parties. "We saw ourselves as anthropologists from the 21st century inhabiting a time module set somewhere in the dark ages of the 1960s," Leary later wrote with characteristic grandiosity. "On this space colony we were attempting to create a new paganism and a new dedication to life as art." The FBI raided Millbrook so often that Leary ultimately shut down the program and turned his attention to a propaganda campaign extolling the virtues of LSD. It was so effective that President Richard Nixon declared Leary a public menace and "the most dangerous man in America." He ended up in Folsom Prison in solitary confinement for several years. In a low point for the First Amendment, the sentencing judge declared, "If he is allowed to travel freely, he will speak publicly about his ideas."

Leary's prison sentence burnished his reputation as a countercultural icon. He was released in 1976, but by then the damage had been done. LSD became a casualty of the culture wars, much to the chagrin of the man who started it all: Albert Hoffmann. While Hoffmann found Leary intelligent and charming, he felt that Leary's delight in provocation about the recreational uses of the drug shifted focus away from bona fide research. He was right. As part of the 1970 Controlled Substance Act, LSD was classified as a Schedule I illegal substance having "no medicinal value." Such a classification makes it virtually impossible to secure federal funding or the drugs needed for trial studies. LSD-related research effectively ceased, and the lingering stigma associated with the drug has hampered scientific progress for the last forty years.

That stigma may now be fading, helped by several studies demonstrating that LSD and its fellow psychedelics are less dangerous and far more medically beneficial than their Schedule I classification would indicate. Yes, LSD is powerful stuff, but tobacco, cocaine, and alcohol produce far more addictive behaviors. While people may be injured while under its influence, there are currently no fatalities on record directly attributable to an overdose of the drug. As for those dreaded flashbacks, they tend to occur primarily in those already prone to psychological problems. And the tales that LSD crystallizes in spinal fluid or in fat cells where it can dislodge and cause nasty flashbacks many years later aren't supported by medical evidence. LSD breaks down completely in the body within hours, and lingering metabolites are purged within a few days. Finally, those early studies touting a link between LSD and long-term neurological damage have not held up to further scrutiny: most had too few subjects and did not control for preexisting mental illness, or the use of other more toxic substances like alcohol or amphetamines.

No drug is entirely without risk. Just check out the list of disclaimers on any given advertisement for many prescription drugs, warnings of possible side effects. Such drugs are prescribed when the benefits for any given user outweigh the risks. Why should psychedelics be any different? "Our whole drug policy is fundamentally mistaken in that it tries to ascribe good or bad qualities to the drugs themselves, and ignores the relationships people have with these drugs," said Rick Doblin, founder of the Multidisciplinary Association for Psychedelic

Studies (MAPS). In other words, it's not the drug that should concern us, but how the drug is used.

Case in point: in the 1920s, a German pharmacologist and toxicologist named Louis Lewin became fascinated by aya-huasca, particularly a compound from the vine that he called banisterine. Lewin ingested it and found that he didn't experi-ence the usual altered states of perception, because the psycho-active compound (DMT) was absent. Rather, he felt strong and vigorous, with increased appetite and improved motor control, and concluded that banisterine might be an excellent treatment for Parkinson's disease, which destroys dopamine receptors in the brain—the same receptors targeted by banisterine and simi-lar compounds.

Banisterine is a monoamine oxidase inhibitor (MAOI), a class of drugs now used widely to stave off the most debilitating effects of Parkinson's. Juan Sanchez-Ramos collaborated with several other scientists in Ecuador on a study. Like Lewin, they found that banisterine improved motor function in those suffer-ing from Parkinson's, even after a single dose.

LSD and psilocybin can be used to treat cluster headaches, severe migraines that can last anywhere from fifteen minutes to three hours. Attacks can occur two to eight times a day in the midst of a cycle. Peter Goadsby, the world's leading researcher on cluster headaches, has called the pain worse than natural childbirth or amputation without anesthetic. Given the severity of the symptoms, and the ineffectiveness of most conventional drugs, chronic sufferers have turned to LSD and magic mush-rooms. One conventional medicine for cluster headaches, meth-ysergide, is chemically similar to LSD.

John Halpern of the Alcohol and Drug Abuse Research Center (ADARC) at McLean Hospital in Belmont, Massachusetts, conducted one of the first scientific studies of this reported efficacy. He interviewed more than fifty sufferers of cluster headaches who admitted to self-treating with psychedelics to alleviate their symptoms. Based on these self-reports, Halpern found that ingesting psilocybin and LSD reduced the pain associated with cluster headaches and could even end the cycles in which they occur. Eighty-five percent said that the drugs aborted attacks, and 52 percent said that the headaches stopped altogether.

The study was admittedly flawed, since there were so many potential sources of bias, but Halpern thought his findings warranted greater study and went on to design a double-blind, placebo-controlled study of psilocybin. While it was still under review, he became intrigued by the promise of a non-hallucinogenic derivative of LSD, BOL-148, first developed in the 1950s. The results from those studies exceeded his expectations: some patients went from chronic headaches to episodic attacks after just one or two doses, and one participant went from suffering forty attacks per week for thirty years to none. In 2010, Halpern founded Entheogen Corporation to develop and bring to market a drug based on BOL-148.

Halpern has also conducted studies on two other psychedelic drugs: one with a Native American church that uses peyote in its ceremonies, and the other with the Santo Daime church in Oregon, which uses ayahuasca. Such use is legal, since it is linked to religious practice. The ADARC is affiliated with Harvard, Timothy Leary's old stomping grounds, but the similarities with Leary end there. Halpern and other scientists interested in the therapeutic properties of hallucinogens show no signs of turning into New Age gurus of any sort. "We want to be the anti-Leary," Halpern told the *New York Times*. "We are serious, sober scientists."

Shakabuku

Well, maybe not completely sober. Halpern admits to ingesting peyote as part of his research. "It would have been extremely insulting if I didn't try it," he said. The Navajo were initially hostile to the notion of an outsider participating in their ceremonies. In that first experience, the "roadman" in charge of his "journey" kept making him ingest more peyote until he vomited. For Halpern, the message was clear: "You want to learn about peyote? I'll teach you about peyote."

The lesson took: it instilled a profound respect for the drug, and for the peyote ceremony as well, which lasts ten hours. Participants chant and pray through the night, and it is not unusual for someone to vomit even though the dosage is small: a few teaspoons, or the equivalent of 100 mg of mescaline. That amount is just enough to amplify the emotions, but not strong enough to bring on full-fledged visions or hallucinations. "This is not a party drug," Halpern insisted, echoing the sentiments of most scientists who study psychedelics. Church members use it as a sacrament for communion with God, not for recreation; the religion strongly discourages the use of other drugs or alcohol.

Halpern's interest in psychedelics was sparked in the early 1990s, when the Indian-born psychiatrist Chunial Roy told him of a survey he'd done among the Indians of British Columbia showing very low rates of alcoholism among the members of churches in which peyote was used. From 1950 to the mid-1960s, academic journals published more than 1,000 research papers on the use of psychedelics to treat some 40,000 patients, including alcoholics. Juan Sanchez-Ramos collaborated on an early clinical study led by Deborah Mash at the University of Miami on the use of ibogaine to treat addiction. Ibogaine is derived from the root bark of an African shrub called *Tabernanth iboga*. Traditionally, the bark is pulverized and swallowed in large amounts, producing visual hallucinations and dissociative states. Yet ibogaine also blocks cravings and alleviates withdrawal symptoms for many opiates, like heroin.

Even the patron saint of sobriety, William Wilson—aka Bill W, founder of Alcoholics Anonymous—tried LSD in 1956 and concluded it could benefit alcoholics by triggering religious experiences like his own, which led him to stop drinking. In the late 1950s, researchers gave LSD to World War II veterans who were chronic alcoholics; when they followed up one year later, they found that 55 percent were still sober. Those results are borne out by a 2011 review by a group of Norwegian graduate students visiting Halpern's laboratory, who pulled together the archived studies of LSD and alcoholism conducted between 1966 and 1970, which involved 536 patients in all. They reanalyzed the data and found that even a single dose of LSD helped heavy alcoholics quit drinking and reduced the risk of relapse. On average, 59 percent of patients showed a clear improvement, compared to 38 percent of patients in control groups.

This is consistent with Halpern's findings that peyote can reduce addicts' cravings for at least two months after a "trip." One Navajo mother told Halpern about her teenage son, who ran away and joined a gang, sinking into a life of hard drugs and alcohol. She didn't hear from him for three years, until he showed up drunk on her doorstep one night and asked for a peyote ceremony. That first experience didn't take: her son became very sick and had to stop, then disappeared for another six months. "Peyote really doesn't like alcohol," Halpern said. But after a stint in detox, the young man decided he wanted to try again. This time, he made it through the entire ceremony, and it changed his life. He sobered up, got a job and an apartment, and started attending night school to learn the Navajo language. There was nothing pleasant or recreational about that young man's peyote experience—that was the point.

"When you take a substance that helps you see how you have been lying to yourself, or how you have automatically assumed certain aspects of reality because it is always processed that way, and then you see it in a different way—when you come back from that experience, well, a Native person will say, 'It sure will lay you flat,'" Halpern said. One Navajo man told him, "If you see lots of pretty colors and geometric objects and all types of goofy things on peyote, you better eat more peyote, because that ain't getting to the heart of the matter."

LSD doesn't much like alcohol either. Late-night talk-show host Craig Ferguson recalled popping some innocent-looking pink pills just before leaving a club one night when he was a hard-partying twenty-one-year-old. He was walking home with a friend through a park when the acid kicked in. "The Victorian statues . . . followed us with their eyes, the wind in the leaves was whispering vague sinister threats, and mysterious ripples bubbled up from the myriad of dark ornamental ponds," he wrote in *American on Purpose*. Already well on his way to chronic alcoholism, Ferguson's anxiety soon escalated into blinding terror, a state in which even a handful of rogue ducks was cause for panic. The experience was so harrowing that Ferguson called his mother from a pay phone. The sound of her voice was enough to pull him back into reality. He never dropped acid again. "Something happened inside my mind that night that should not have happened," Ferguson wrote. "Acid gave me a clinical, unblinking look at madness, and I discovered I wasn't brave enough to be insane."

Halpern would argue that this is exactly what should happen to a clinical alcoholic on acid. Most alcoholics hate hallucinogens, he has found, because the effects are the opposite of what they are trying to achieve when they drink. They are trying to escape—from stress, a fight with their spouse, their own anxiety or depression—and thus need a predictable substance that will give them a familiar feeling of comfort and help them cope, at least for a little while. Psychedelics are predictably unpredictable. They focus your thoughts inward and have a way of dredging up everything an alcoholic is most trying to avoid. "Psychedelics aren't good for escapism; they're going to make you think," said Robin Carhart-Harris. "Instead of allowing users to avoid negative emotions, they magnify the painful feelings. This may help patients address their problems instead of fleeing them, but it can also exacerbate distress."

So why are psychedelics so effective in treating alcoholism? David Nichols, a pioneer in the study of Ecstasy (MDMA), compared the compulsive behaviors of alcoholics and other addicts to a personal computer that has become caught in a programming loop and needs to be rebooted. The brain can become caught in destructive loops too. "There are behavioral subroutines or programs that accumulate over the years and produce dysfunctional behaviors," he explained. "You are not aware of them. They are below the level of consciousness." LSD and other psychedelics can literally reboot the brain and make it easier to reset some of those destructive subroutines. It is akin to the Buddhist concept of *shakabuku*, defined by Debi, the hometown DJ in the film *Grosse Pointe Blank*, as "a swift, spiritual kick to the head that alters your reality forever." There is evidence that a single dose of psilocybin can lead to pronounced changes in personality, most notably in the openness factor, which relates to creativity, fantasy, emotions, and empathy. Those changes persist even a year after the dosage.

That is why a handful of scientists are revisiting the earlier research using hallucinogens to treat anxiety and depression that was so abruptly terminated back in 1970. Nichols's Brazilian-born wife, Cibele Ruas, is a psychologist who estimates that around 75 percent of all psychological disorders are linked in some way to multiple forms of anxiety and depression. Any drug that could address three-quarters of cases in clinical psychology would be extraordinarily valuable. A good place to start might be using psychedelics to help terminally ill patients accept the reality of their own death. "You can look at anxiety concerning death as the blueprint for anxiety in general," said Ruas. "Because fear of death is the first and ultimate fear."

Halpern has administered MDMA (the compound, not recreational Ecstasy) to terminal cancer patients, as has Charles Grob, a psychiatrist at Harbor-UCLA Medical Center. There is less of a stigma attached to MDMA and psilocybin: they are easier to acquire, equally safe, and nontoxic. Their chemical properties are similar to LSD, as are their psychological effects. Roland Griffiths, a behavioral biologist at Johns Hopkins University, has found that a dose of 20 to 30 mg of psilocybin can induce the equivalent of a profound spiritual experience in many people. His subjects reported feeling the boundary between self and other dissolve, becoming part of some larger

state of consciousness, in which their personal worries and inse-
curities seemed trivial. Those peaceful feelings persisted even
fourteen months later, when 94 percent still described the experience as one of the five most meaningful of their lives. They
also reported feeling less fear and anxiety about death.

One of the most moving accounts of this effect appears in a
letter written by Laura Huxley shortly after the death of her
husband. Aldous Huxley was diagnosed with throat cancer in
1960. By November 1963, he did not have much longer to live.
Huxley was in considerable pain and his doctors warned that
the end would not be pretty: most people with late-stage throat
cancer died in violent convulsions and fits.

Wishing to ease his final moments, both physically and psychologically, Laura asked her psychiatrist, Sidney Coleman,
whether the drug could prepare Huxley for his impending
demise. It was not unprecedented to give LSD to dying cancer
patients. Czech-born psychiatrist Stanislav Grof had done so
for several in his care at the Spring Grove State Hospital near
Baltimore, with promising results. But Coleman had prescribed
LSD only twice; in one case it proved beneficial, and in the other
it made little difference, so his results were inconclusive.

In the end, it was Huxley who made the decision. On the
morning of November 22, he scrawled instructions for Laura on
a sheet of paper: "LSD—try it—intramuscular—100 mg." Huxley's physician had misgivings, but shrugged them off: "At this
point, what is the difference?" Laura herself administered the
injection. She gave Huxley a second 100 mg dose an hour later,
leaning close and whispering encouraging words to him for the
next four hours.

As the afternoon waned and reports of President John F.

Kennedy's assassination dominated the news, Laura recalled, "The twitching stopped, the breathing became slower and slower, and there was absolutely not the slightest indication of contraction or struggle." Huxley died at 5:20 p.m. Instead of being marked by violent convulsions, his passing was "like a piece of music just finishing so gently *in sempre piu piano dolcemente*. Both doctors and nurse said they had never seen a person in similar physical condition going off so completely without pain and without struggle." She ended her letter with a sobering question: "Now, is his way of dying to remain our, and only our, relief and consolation, or should others also benefit from it?"

Laura Huxley's account jibes with my own experience. I felt an odd sense of peace and comfort at being enveloped in a vast universal architecture, making my inevitable demise seem far less fraught with significance. Yet even at the height of the trip, when I was merely a disembodied collection of molecules, floating randomly in some universal neurospace, there was still a sense of what some psychologists call a "militant I" lurking in the background. That persistent core self is one reason why some neuroscientists suspect the ways in which LSD affects the brain might one day provide crucial keys to decoding the mystery of human consciousness.

Cibele Ruas compared consciousness to a fenced-in yard, where something akin to Huxley's "reducing valve" imposes mental limits on infinite possibilities. "If you tear up the fence, you face the whole universe, and that is too much. You cannot function," she said. Psychedelic drugs like LSD remove those constraints temporarily. "You are playing in a dangerous terrain" when you take acid, said Ruas. "You are playing outside of the box." Whether you find the experience frightening or exhilarating may depend in part on your disposition. LSD is powerful stuff, and it isn't for everyone.

As the effects of the drug wore off, my self began to reconfigure, almost as if I were waking from a dream. By nightfall, Wonderland had all but disappeared. But when I closed my eyes, I could still detect faint wisps of colorful curlicues—a gentle reminder that "reality" is not always what it seems.

8

Ghost in the Machine

I wake to sleep, and take my waking slow.
I feel my fate in what I cannot fear.
—THEODORE ROETHKE, "THE WAKING"

It was the wee hours of the morning, and I should have been slumbering peacefully in my cramped East Village studio. Instead, I was standing alone in a dingy room in the ICU at Beekman Downtown Hospital, blinking back tears as I confronted an emaciated, too-still figure lying on a bed under harsh fluorescent lights. I had seen dead bodies before at open-casket funerals, embalmed and painted like life-sized waxen dolls, but this was different. Nick had been dead only thirty minutes.

His disease-ravaged body was still warm to the touch as I took his lifeless hand in mine and tried not to look at the bony clavicle jutting through the hospital gown or the yellowed and rusty stains on the sheets. I could feel the heat seeping away from his body; the metabolic engine had stopped. Sleep is often poetically declared to be the mirror image of death, but Nick looked nothing like a man in deep slumber. There was no telltale rise and fall of the chest, his open eyes stared vacantly at the ceiling, and a bit of dried spittle lingered at the edge of his slackened mouth. I touched his stubbled, sunken cheek and tried to gently close the eyelids, as I'd seen so many people do in films, but they just popped open again. This thing on the bed couldn't be Nick; it was a loose collection of skin, sinew, and bone that just happened to wear his face. Every other trace of the man I knew was gone.

More than twenty years later, I still remember that moment as vividly as if it happened yesterday. There was nothing unique about my reaction; I reacted the same way human beings have responded to the dead bodies of loved ones for millennia: *Where did they go?* The physical body is present, made up of the same atoms that make up everything else in the material world; there is still some*thing* there. It's the some*one* that has gone. We have a powerful sense of a mysterious, intangible quality that is missing—what the ancients called an anima ("the spark of life"), and philosophers and theologians came to term the soul. In 1907, a physician in Massachusetts named Duncan MacDougall even tried to weigh the soul, using six elderly patients in the final stages of tuberculosis. When he deemed death was near, he placed the patient on an industrial scale and looked for any drop in weight coinciding with the moment of death. The soul, he determined, weighed about twenty-one grams. MacDougall is dismissed as bit of a crackpot in the footnotes of science history—his results are highly suspect—but the notion that there is an animating substance separate from the physical body is a persistent one.

From a neuroscientific perspective, "soul" is synonymous with "mind," a broad term that encompasses perception, memory retrieval, subconscious processes, and aspects of consciousness—the most fundamental aspect of the self. For a neuroscientist, it is a question of whether a certain type of brain activity is present. In that regard, our intuitive sense of a ghost in the machine is incorrect. But we are still a long way off from fully understanding the mechanisms underlying the fully conscious self. Until quite recently, only a handful of scientists bothered to study the topic at all. In 1996, a curmudgeonly British psychologist named Stuart Sutherland—charged with providing entries for the second edition of the *International Dictionary of Psychology*—defined consciousness thusly: "A fascinating but elusive phenomenon; it is impossible to specify what it is, what it does, or why it evolved. Nothing worth reading has been written about it."*

Sutherland's sweeping dismissal is arguably unfair; most neuroscientists today agree that the mind *is* the brain. This is an area where science bumps up against philosophy, and it is all too easy to become hopelessly muddled while trying to sort out the science from the speculation. A little expert guidance was in order.

In search of that guidance, I found myself in a large ranch house just north of San Diego, seated at Patricia Churchland's kitchen table, chatting about consciousness and the self over tea. Occasionally one of her golden retrievers nudged my lap for a round of petting—bona fide attention hounds—and Churchland would let me indulge the dogs briefly before shooing them away. She and her husband, fellow philosopher Paul Churchland, have carved out illustrious careers grappling with language, morality, ethics, and the relationship between consciousness and the brain. "I'm quite bullish on conscious-

---

*In the same volume, Sutherland defined love as "a form of mental illness not yet recognized by any of the standard diagnostic manuals."

ness right now," Churchland said with characteristic frankness. "Ten years ago there were a lot of people talking about it and they didn't have anything to say. There are still a lot of people who have nothing to say, but there are also a handful pursuing it with some really good ideas, not just waving their hands."

Churchland has a knack for cutting through the crap, and she wasted no time setting me straight on what she deems most fundamental. "People have focused on 'fancy consciousness,' where we can talk and self-reflect and engage in metacognition," she explained—what might better be termed self-awareness. "We think that's what consciousness is all about, but it's not." Churchland suggested recasting the discussion of consciousness in different terms. "How rich is your self-representation? That is the real question."

Consider C. *elegans*. The nematode has a mere 302 neurons, barely sufficient to qualify as a functioning brain, and part of one of those neurons is devoted to distinguishing its own body from the world around it. C. *elegans* has a "protoself"—that is, it can distinguish between its own movement and movement in its environment. In the mid-nineteenth century, Hermann von Helmholtz insisted that all animals needed this basic ability to discern self from other, and he offered a simple experiment to demonstrate this. If you gently press on your own eye, it will seem as if the world in front of you has moved, when in reality the image is moving across the retina. Whenever it issues a motor command, the brain creates something called an "efference copy"—a parallel simulation of the predicted movement and its expected sensory results, which it then matches up with the eventual outcome. The movement of the eyeball is passive, in contrast to when the eye muscles move the eyeball; then, the image of the world will appear to be still. Because the passive movement of the eyeball doesn't register with the brain's visual system, it affects your ability to distinguish the world's motion (or lack thereof) from that of your own body.

"I think it's fair to say that C. *elegans* has a very primitive self-representation," said Churchland. "If it can distinguish between its own movement and other movement, that is a self." This does not mean that the nematode has a conscious sense of self on a par with human beings, although it's not like we can ask C. *elegans* directly. How do we know the little creature doesn't have a rich interior life, arguing with its hipster friends about that scathing film review in the latest issue of *Nematoda Cineaste* over tiny cups of espresso in roundworm cafés? Let's just say this possibility is highly unlikely, given what we know about its neuroanatomy. It is the number of neurons that matters most when it comes to the richness of self-representation, and the nematode doesn't have that many.

Yet we do share some basic neural structures and functions with many animals, a fact that led a number of leading neuroscientists to sign a public statement in August 2012. The so-called Cambridge Declaration on Consciousness concluded that, while not self-aware in the human sense, "non-human animals"— including mammals, certain insects, and the octopus—possess sufficient brain structure and function to generate conscious states. One reason Churchland is more bullish on consciousness research today than she was ten years ago is because neuroscientists are homing in on the underlying neural structures and processes that give rise to consciousness. It turns out that the poets were partially correct: consciousness appears to be linked to sleep-wake cycles.

No Matter, No Mind

The seventeenth-century philosopher René Descartes was the personal tutor of Queen Christina of Sweden, but it was an exiled princess who notoriously bested him in a philosophical debate. Princess Elisabeth of Bohemia was the daughter of King Frederick V and the third of thirteen children. She had very little time to enjoy her royal status, since her father was ousted when she was barely a year old. The family fled to Holland, where the princess received much of her education, becoming so proficient in languages, math, and the sciences that her siblings dubbed her "La Greque." Among the philosophical works she read were those of Descartes, who espoused the notion of mind-body dualism—namely, that the two are separate entities. The body, he reasoned, was a physical machine, existing in the material realm and therefore subject to the laws of nature. The mind, or soul, he deemed nonmaterial—an essence that transcended the material realm, the ghost in the machine. Hence Descartes's famous dictum, "I think, therefore I am."

Elisabeth took issue with this concept, arguing that if mind and body were distinct and separate entities, they would not be able to interact, as Descartes claimed they could. (He thought the mind controlled the body, although the body could affect the mind in turn, as when one's reason was overcome by passion.) The two fired letters back and forth for the next seven years until the philosopher's death in 1650, and while some historians have speculated that their relationship was more than epistolary, there is no conclusive evidence to that effect. Descartes never did come up with a satisfactory answer for his princess, and philosophers continued to debate the mind-body problem for the next several centuries. Descartes even pin-pointed what he believed to be the seat of the soul: the pineal gland—a tiny pea-shaped region near the center of the brain that secretes the hormone melatonin, which is tied to sleep-wake cycles.

Scientists have since ruled out the pineal gland as the seat of the soul, but 350 years later, the Nobel Prize–winning biologist Francis Crick—co-discoverer with James Watson of the helical structure of DNA—found a new potential culprit. Crick posed his own question: what is the most fundamental aspect of con-sciousness? He reasoned that it was the sense of continuity and unity to our sense of self, despite being comprised of so many different subcomponents. He zeroed in on a thin layer of tissue just under the brain's insular cortex as a possible source of this unity. Known as the claustrum, this region connects to nearly every part of the brain, both sending and receiving signals continuously.

Another suspect is the thalamus, sandwiched between the cerebral cortex and the midbrain. Patients in vegetative states usually have an atrophied thalamus, as well as damage to the white-matter tracts that carry nerve signals to and from that region. Irreparable damage to the thalamus was at the heart of the controversial Karen Ann Quinlan case in the 1970s. Quin-lan suffered a heart attack at twenty-one after mixing Valium (diazepam), painkillers (dextropropoxyphene), and alcohol at a party. The twenty minutes of respiratory failure she suffered left her in a vegetative state. Her family sued for the right to pull the plug on the elaborate life-support system that was keeping her alive—if one could call that living. Eventually what was left of Quinlan died of complications from pneumonia, and the autopsy revealed that while her cerebral cortex and brain stem were largely undamaged, her thalamus had been destroyed.

But the thalamus does not act alone, given the distributed nature of the brain's neural network. Patricia Churchland sketched a rudimentary diagram of a three-pronged structure on a scrap of paper: the brain stem–thalamus–cortex axis.* The brain stem connects to the thalamus, and the thalamus projects to the cortex, and the cortex projects back, in a continuous feed-back loop. Damage or sever those connections, David Eagleman explained to *Science News*, and "that individual will not be aware by any measure, forever."

Nicholas Schiff, a neuroscientist at Weill Cornell Medical College, has studied the implications of the brain stem–hypothalamus–cortex axis for consciousness, based on work with coma patients. One such patient was a fireman who had been a coma for nine years. Schiff could find no evidence of damage to the cortex or brain stem; it was if the fireman was in a deep sleep. One morning they included a dose of the sleeping pill Ambien in the fireman's medication cocktail. The man woke up and immediately asked, "Where am I?" This wasn't the first time this has happened. The first case appeared in 1999, when a South African man was hit by a truck and lapsed into a vegetative state. Doctors noticed him clawing at the mattress during the night, a symptom of insomnia, and gave him a drug called zolpidem (the generic name for Ambien) to calm him down. To their shock, the man woke up, fully conscious and able to speak a few words and control his limbs and facial

---

*This axis pertains to what philosopher Ned Block calls "access con-sciousness," as opposed to "phenomenal consciousness"—how we experi-ence the color red or the sound of a clarinet, for example, or our emotions.

muscles. When the drug wore off a few hours later, he lapsed back into a vegetative state. His doctors kept administering zolpidem each day to revive him, a few hours at a time, and eventually he no longer needed the drug to stay awake and conscious.

Within the brain stem, there are objects called intralaminar nuclei that help regulate sleep-wakefulness cycles, as well as arousal, attention, and emotions. When we are awake, those neurons fire many times per second, like a brainy biological clock. During sleep—but not during dreaming—this system shuts down, a true loss of consciousness. Zolpidem, propofol, and similar drugs jump-start those nuclei, just like a car battery, so that the normal sleep-wake cycle kicks in.

The basal ganglia also get involved. This is a group of brain nuclei underneath the front of the brain that controls simple motor tasks and habits that we perform without conscious thought. But whether we realize it or not, the basal ganglia are whirring away beneath the surface, making decisions via two feedback loops. One loop serves as a brake, or an "off" switch, stopping someone from physically acting out his or her dreams. We have all had that experience of trying to run or fly or punch an opponent in our dreams, only to feel frustrated because we just couldn't do it. Blame your basal ganglia—or thank them, since otherwise we all might be running in place and punching our partners in our sleep. The second loop releases that brake— it serves as the "on" switch—and it's this loop that seems to be affected by sleep-aid drugs like Ambien and propofol. They trigger a reaction called paradoxical excitation. Though a common side effect, it eventually leads to deeper sleep. People on Ambien have been known to walk or eat while still asleep.

Consciousness really doesn't flick on and off. It works more like a dimmer switch. Brain regions don't all shut down at the same time when you lose consciousness, nor do they switch back on simultaneously when you emerge from general anesthesia, which Schiff likens to a reversible drug-induced coma. He monitored the electrical activity in the brains of patients under general anesthesia (propofol and dexmedetomidine), noting the stages in which the brain regions "woke up." The cerebral cortex was not the first region of the brain to reemerge from unconsciousness. Instead, it was the most primitive core structures of the brain, specifically the thalamus and parts of the limbic system. As the drug wore off, the most basic functions reemerged first: those that control breathing, then salivation and tear ducts, followed by swallowing and coughing. Finally, the patients' eyes fluttered open and they began to notice and respond to the world around them.

But does this really have much to do with the self? "Consciousness is not merely wakefulness," Antonio Damasio, a neuroscientist at the University of Southern California, wrote in *Self Comes to Mind*. "There is indeed a self, but it is a process, not a thing, and the process is present at all times when we are presumed to be conscious." So the conscious self isn't located in any particular part of the brain; rather, it is what's known as an emergent phenomenon.

Emergence can be difficult to define, but in essence, it describes a system in which the whole is greater than the sum of its parts—a concept that dates back to Aristotle.* Consider a lump of gold, made up of atoms that together comprise a complex network. Gold is a substance with distinct physical properties,

---

* "The totality is not, as it were, a mere heap, but the whole is something besides the parts."—*Metaphysics,* Book H.

such as temperature, conductivity, and color, but those properties are not found within the individual atoms. Rather, they
emerge from the many different interactions between those atoms. Similarly, consciousness is a property that emerges from
the many interactions between networked neurons in the brain.
That doesn't mean the conscious self is illusory. Temperature
and conductivity are real, measurable properties. A traffic jam
is emergent too, but nobody would dismiss it as a mere illusion.

That is the reductionist answer to Descartes's mind-body
problem: mind *is* matter, and consciousness is emergent.* "No
matter, no mind," Caltech neuroscientist Christof Koch shrugged
matter-of-factly. "If there is no matter, my mind doesn't exist."
This thing we call the soul is something unique, generated, he
insisted, "by the causal interaction of myriad elements in my
head. But once this brain dissolves in death—or at night when I
go into a deep sleep, or go under anesthesia—my consciousness
flees." Neuroscientists have been less successful when it comes to
teasing out the specific mechanisms behind how those processes
produce a distinct, persistent, subjective personal identity—
Churchland's "fancy consciousness." On one end of the scale,
there is basic wakefulness and sentient awareness of one's environment; on the other, there is self-awareness. That is why Damasio draws a critical distinction between the Self-as-Object (the
material "me") and the Self-as-Knower (the subjective, self-
aware "I"), although he emphasizes that the two are inextricably

---

*There is an ongoing debate in philosophical circles regarding "weak"
versus "strong" emergence. I am focusing on weak emergence, in which a
new property arises from interactions among individual components of a
system. Proponents of strong emergence would contend that this new
property is irreducible to its component parts.

linked. The Self-as-Object is that fundamental layer we seem to share with animals, onto which human beings add a second, richer layer of self-representation, the Self-as-Knower.

Remembrance of Things Past

One evening in 1671, a group of philosophically minded friends gathered at the London home of Lord Anthony Ashley Cooper, first Earl of Shaftesbury. Among those present was John Locke, a physician under the earl's patronage who had saved the aristocrat from a near fatal liver infection. Locke had an interest in philosophy as well medicine, and while his account of that meeting is scant on specifics, he wrote that the conversation that night made him realize "it was necessary to examine our own abilities and see what objects our understandings were, or were not, fitted to deal with."

Thus was born Locke's famous *Essay Concerning Human Understanding*, his attempt to grapple with how human beings think and perceive and construct a self. In Book II, Locke maintained that memories were key to the construction of our personal identities. Known as episodic or autobiographical memories, we string these past experiences and actions together over our lifetimes to form a continuous, coherent personal narrative. Someone who had no memory of his or her past would, in essence, have no self.

It was an intriguing argument, and Locke's ideas have influenced many philosophers since the *Essay* was published in 1690. Unfortunately for Locke's legacy, neuroscientists have learned a great deal about how memory works in the interim, and the evidence suggests that autobiographical memory—a coherent personal narrative—may not be quite as fundamental to the self

as Locke thought. Even someone with little or no autobiographical memory still retains some semblance of an "I."

There is no more famous case in memory research than the
patient known as H. M. Born in 1926, H. M. suffered from severe epilepsy, possibly as the result of a bicycle accident when he
was nine, although the condition ran in his family. By the time
he was in his mid-twenties, he suffered ten seizures a week and
proved unresponsive to any of the available drugs at that time.
So in 1953, the neurosurgeon William Beecher Scoville—a
colleague of neurosurgery pioneer Wilder Penfield—performed
a radical surgery that removed large chunks of brain tissue from
H. M.'s left and right temporal lobes, including the amygdala
and most of the hippocampus. The surgery worked, kind of:
from then on, H. M.'s seizures occurred only a couple of times
a year and were not as severe. But the "cure" came at great personal cost. He lost his ability to form new long-term memories,
unable to retain any new information for more than a few minutes. Psychologist Brenda Milner, who treated H. M. for thirty
years, noted in a landmark 1957 paper that after the operation

> this young man could no longer recognize the hospital
> staff, nor find his way to the bathroom, and he seemed to
> recall nothing of the day-to-day events of his hospital
> life. . . . [H]e did not remember the death of a favourite
> uncle three years previously, yet could recall some trivial
> events that had occurred just before his admission. . . .
> His early memories were apparently vivid and intact.

H. M. wasn't completely incapacitated. He could learn simple skills with practice, such as tracing a diagram. He could understand jokes, remember words, and recall and draw the floor

plan of the house he had lived in for eight years. But he couldn't draw the floor plan of his current residence, and while he could learn new motor tasks, he had to be reminded how to perform them. He never once recognized Milner during thirty years of treatment. This is why she famously described him as being "chained to the past." H. M. died in 2008.

The problem lay not with H. M.'s short-term memory, but with his ability to transfer new memories to long-term storage. This transfer usually happens during the REM (rapid eye movement) sleep cycle, which is also when most of our dreaming takes place, as our brain processes the day's events, sorting through the new information and integrating it with older memories to maintain a cohesive whole. "We know that memory consolidation does not happen without you going offline," said Patricia Churchland. "The hippocampus is teaching the cortex [when we sleep]."

In 1991, an MIT neuroscientist named Matt Wilson was recording neural signals in rats as they ran a maze in the laboratory, using an ingenious system of implanted electrodes. One day, he left the rats hooked up after they had finished the maze and fallen asleep, while he analyzed the day's raw data. He became aware that he was hearing unusual neural firing patterns from the brains of the slumbering rats—patterns remarkably similar to those he recorded as the rodents were awake and running the maze. Their brains were replaying that earlier activity, reinforcing what they had learned and transferring it to long-term memory. The neural firing patterns were so similar, Wilson realized the rats had to be retracing the path through the maze they had taken that day. Human beings do this too. That is why it's best to study for an exam the night before and then get a good night's sleep, to ensure that long-term memory and learning takes place.

H. M. still had his past personal memories and his sense of self, so was Locke correct in ascribing so much importance to autobiographical memory? This seems unlikely, primarily because of another famous patient called Boswell, who suffered far more severe memory loss than H. M. Thanks to a brain infection at age forty-eight, Boswell lost almost all of his autobiographical memories. He couldn't form new memories or recall anything about his personal life, including the existence of his wife and children. But he could play checkers (although he called it "bingo"), he had personality traits and social skills, and he was clearly conscious—in short, he still had a self, right down to the correct use of the "I" pronoun when talking about himself. For Churchland, this is powerful evidence that Locke was wrong: memory is not the key to our sense of self, because if it were, Boswell wouldn't have a self.

Autobiographical memory is certainly important to identity; it's just not central to consciousness. Rather, Churchland explained, our sense of continuity, of a unified whole, occurs because all this processing is happening inside a single brain.

Much of our neural activity occurs below the conscious level. Incoming sensory information is processed in specialized brain regions, but "we" are not aware of that input immediately. It is only when those processed signals are transmitted through the networks of neurons in other regions of the brain—most notably the cortex, the seat of higher brain functions and integration—that we become conscious of the experience. We lose consciousness when this process is disrupted.

A 2011 study of volunteers under propofol found that as the

anesthetic took effect, certain brain regions that normally worked together fell out of sync; they ceased to communicate. When the patients were unconscious, small sections of the sensory cortex still fired in response to outside stimuli, but those signals did not spread to other areas. "Long-distance communication seems to be blocked," neuroscientist Andreas Engel told *New Scientist*. "It's like the message is reaching the mailbox, but nobody is picking it up." When the patients were awake, or only mildly sedated, that communication was not disrupted.

"The patterns that anesthesiologists see support the notion that consciousness emerges from the integration of information across large networks," Schiff's colleague, anesthesiologist Emery Brown, told *Technology Review*. Current research points to a wide distribution across brain structures that coordinate with one another only when needed—a network of networks that Churchland described as a "loose and loopy hierarchy"—and it is this that gives rise to conscious awareness. So how does that activity become so widespread in such a short time? Simply having a large number of connections or interactions among the components of the brain isn't sufficient. Consciousness also depends on how those networks are organized.

It's a Small World

Sometimes inspiration arises from sheer boredom. During a snowstorm in 1994, three students at Albright College found themselves watching *Footloose*, a film starring actor Kevin Bacon as a big-city boy with terrific rhythm who is stuck in a puritanical small town where dancing is forbidden. As soon as *Footloose* ended, the same station aired *The Air Up There*, also starring Bacon. The young men wondered how many films Bacon had been in, and by extension, how many other actors could be connected to him through those films. Thus was born the trivia game Six Degrees of Kevin Bacon, based on the "small world" phenomenon. The premise is that any actor can be linked through his or her film roles to Bacon within six steps.*

The higher someone's "Bacon number," the further they are removed from direct association with the actor. Kevin Bacon has a Bacon number of 0. Those who worked directly with him have a Bacon number of 1, and those twice removed have a Bacon number of 2. This was demonstrated in a 2002 Super Bowl commercial for the Visa check card featuring Bacon trying to write a check in a bookstore—except he forgot his ID. To verify his identity, he corrals a group of people to present to the clerk:

> OK. I was in a movie with an extra, Eunice, whose hairdresser, Wayne, attended Sunday school with Father O'Neill, who plays racquetball with Dr. Sanjay, who recently removed the appendix of Kim, who dumped you sophomore year. So you see, we're practically brothers.

Eunice has a Bacon number of 1, Wayne is a 2, Father O'Neill is a 3, Dr. Sanjay is a 4, and Kim is a 5; the highest Bacon number yet reported is an 8.† So is Bacon really the center of Hollywood's universe? He is highly connected, a hub in the in-

---

*Bacon wasn't thrilled when he first heard about the game, but soon recognized the humor and embraced the concept, even establishing a charity in 2007 called SixDegrees.org.

†Traditionally, the Bacon number only refers to actors.

dustry, but according to the online Oracle of Bacon—a computer algorithm that scans the Internet Movie Database and determines the degrees of separation for each given actor—he doesn't even crack the top one hundred. The rankings are constantly in flux, as new films are released, but top-ranking hubs have included Rod Steiger, Donald Sutherland, and Dennis Hopper.

Small-world characteristics can be found in systems as diverse as road maps, airline transportation, food chains, electric power grids, and social networking sites like Facebook and Twitter. Why is this organizational structure so common to complex networks? A breakthrough paper published in 1998 by Duncan Watts and Steven Strogatz shed some light on this via a simple lattice (honeycomb) model they devised on the computer. The model network was "highly clustered": each individual node, or cell in the honeycomb, was connected to each of its four nearest neighbors, by either a line or an edge, but there were no long-range connections with distant nodes. All those short-range connections fostered good communication flow between neighboring nodes, but not with nodes cast further afield.

The culprit is a property called minimum path length, a measure of the network's efficiency. It is calculated by taking the average of path lengths between all possible pairs of nodes to determine how long it might take for a signal to travel from a node on one side of the network to one on the other side. In the regular lattice model, you have many local node connections with short path lengths, but the path lengths between pairs of distant nodes are very long. Average all that out, and you get a minimum path length that is also very long. This makes it very inefficient and time-consuming to transmit a signal across a network, because it has to pass through many short-range connections before it can reach its goal. Any New Yorker who has taken the local 6 train from downtown to the Bronx during rush hour has experienced frustration at the train's slow progress, because it must stop at every single station. Those lucky enough to squeeze onto the jam-packed express train can cut their commute time in half.

Watts and Strogatz next did a bit of creative rewiring, replacing just a few of the short-range connections in their computer model with long-range connections to nodes that were further afield. This didn't affect the local clustering: you still had those little cliques. But it had a huge effect on the network's overall efficiency, cutting the minimum path length significantly. Now a signal didn't have to hop from one neighboring node to the next. It could take the express train instead of the local, thereby reaching its destination much more quickly.

Small-world networks hit the sweet spot between being too regular and too random, and the payoffs are immense. The U.S. airline industry boasts the quintessential small-world network. We all moan about the inconvenience of connecting flights, but we can usually get from any city to another in two flights—maybe three for especially remote towns. That is because there are a lot of local connections and a few major hubs providing those critical longer connections. In the end, your minimum path length is greatly reduced. Imagine what a pain it would be if you had to hop from one local airport to the next to get from Los Angeles to New York. Alternatively, if there were only major hubs, you couldn't fly to smaller towns at all.

Your brain also makes good use of the small-world model. This is true whether we're talking about the connections between individual neurons, or about how those connections form networks that synchronize with one another as needed to perform various functions. Even large-scale neural networks like the visual system or the brain stem exhibit small-world characteristics. In order to function efficiently, the brain must have local specialized "cliques" to process various kinds of sensory stimuli, but it also needs to be able to distribute and integrate that information broadly throughout the brain.

Somewhere along the evolutionary chain, Mother Nature hit upon small-world organization to handle information processing in the brain; it shows up in several species. *C. elegans* is one of the simplest neural systems, with 2,462 synaptic connections, and yet its architecture follows the small-world model. In 2010, scientists from the University of Notre Dame analyzed seventy years' worth of anatomical data on macaque brains and found that the small-world structure proved consistent across many samples. The number of connections was largest between those areas of the brain closest to each other (local clustering) and declined as distance increased. But there were also a few long-distance connections, acting much like control switches to coordinate how information was exchanged across the entire neural network.

Giulio Tononi, a psychiatrist and neuroscientist at the University of Wisconsin–Madison, has suggested that consciousness is really a form of integrated information, and he devised a metric for determining how much information is in a given neural network—a means of quantifying consciousness. He calls it *phi*. Think of how a photodiode works. Photodiodes respond to incoming light with electric bursts of activity. This translates into low information, because while photodiodes can determine whether it is light or dark, shades of gray or the hues of the rainbow are beyond their powers of detection. In techno-speak, a photodiode can only be in one of two states at a time. Construct a gigantic array of photodiodes and you still wouldn't have a unified consciousness; each is independent from the others and there is no communication among them. Neurons, in contrast, collectively contain a great deal of information, because of all those critical interconnections. The brain can be in any one of trillions of states, which makes calculating *phi* humanly impossible. Even doing so just for *C. elegans* and its paltry 302 neurons would take the lifetime of the universe.

It is not sufficient for the brain's various overlapping networks to receive and process incoming sensory information; for consciousness to emerge, that information must be integrated. A group of disconnected neurons will have a very low *phi*, since they cannot share information. Then again, if every single neuron is linked to every other neuron, *phi* is also low—the system loses its ability to discriminate and becomes one gigantic photodiode, which can only be in an on or off state. That is what happens in an epileptic seizure: many neurons turn on and off at the same time, reducing the number of possible brain states, thereby lowering the system's *phi*. Tonini found that the best way to optimize the neural system is to have parts that are organized into separate clusters and then joined up—essentially a small-world organization. This makes *phi* "a precise measure capturing the essence of consciousness," Christof Koch observed. "The larger the *phi*, the richer the conscious experience of that system."

Francis Crick shocked colleagues by devoting the last two decades of his scientific career to decoding the mystery of human consciousness. He underestimated the magnitude of the challenge. "I think it was a great disappointment to Francis that he did not make a great deal of progress on this problem," said Patricia Churchland. "I think he thought that the paradigm of finding the helical structure of DNA might apply also to consciousness, that if we could just find the structural correlate, then you could pull on that string and everything else would come. And I don't think that's going to happen."

So is the small-world model for the human brain the equivalent of the discovery of the double helix in biology? "That's an interesting question. I wouldn't know," she mused, adding that compared to physics, neuroscience is still in its infancy, teetering on the edge of its own Copernican revolution. "We're pre-Newton, pre-Kepler. We're still sussing out that there are moons around Jupiter."

A World Without "I"

Death approaches so slowly for someone with AIDS that the end, when it arrives, can seem downright sudden. Nick's illness was a steady decline, as he battled several bouts of *pneumocystis* pneumonia, a case of thrush, and countless other bizarre infections. Most disturbing was the onset of dementia, because it changed his personality. Gone were the sardonic humor and staunch pragmatism, replaced by an almost mawkish sentimentality and mostly harmless delusions. Late one night he showed up on my doorstep in the custody of two policemen. They had found him wandering the terminal at LaGuardia Airport with all his high school swim-team medals around his neck, convinced he was catching a flight to Los Angeles to compete in the Special Olympics. My address was the only one he could remember. His doctors prescribed a new drug regimen that reversed the dementia and stabilized the decline—at least temporarily. But he was never quite the same after that, as if something within had been irretrievably broken.

Perhaps something had been broken. Small-world networks are robust because their survival doesn't depend on any given node. Yet for all their efficiency, they can evince the same steady decline. There is a critical threshold to that decline that leads to sudden catastrophic collapse—an effect that is especially pronounced when you have a network of networks. Such a system can absorb the loss of a few local nodes here and there, but knock out enough of those crucial hubs and a pattern of cascading failure kicks in. We all know what happens to the airlines' on-time flight schedules when a snowstorm grounds all the flights at Chicago's O'Hare Airport; the effect ripples out to other airports around the world, delaying flights in cities where there are no adverse weather conditions at all.

"It's as if someone threw a switch, but there is no switch," Boston University physicist H. Eugene Stanley told *Science News*. He thinks this might also be applicable to sudden deterioration in a patient's health. "Breaking a hip could trigger a series of disconnections in a body's network of networks. It is widely known that an elderly person who fractures a hip faces a greatly increased chance of dying within the next year even if repair surgery is successful."

As with the body, so with the brain. The brain is surprisingly adaptive and robust; it can absorb the loss of a few neurons here and there, and sometimes it can even rewire itself to compensate for more serious damage. But long-term damage, such as that wreaked by Alzheimer's disease, leads to serious (and heartbreaking) cognitive deficiencies as the network breaks apart bit by bit. Eventually, the body and the brain's network of networks fail catastrophically.

Nick collapsed suddenly one Friday night. He had seemed so stable just a few days before. But all those interconnected networks in his body finally hit that critical threshold and the cas-cading failure mode kicked in. He hung on long enough to give his friends a chance to say good-bye, before quietly slipping away in the dead of night when nobody was looking.

This is what I realized that gut-wrenching night when I found myself alone with Nick's lifeless body: something essen-tial does depart. There is more to death than just the shutdown of the body's metabolic engine; the brain shuts down too, and once that happens, the self evaporates, because human con-sciousness is emergent. It is all those underlying processes, the constant flow of neural information, that give rise to conscious-ness, which is why significant disruptions in that flow lead to unconsciousness. Once those processes cease entirely, the self disappears forever.

That non-being lies at the heart of the all-too-human fear of death. From the moment we are born, we experience the world through the individual lens of our conscious self. But when we face the prospect of death, we must confront the knowledge that the world will go on without us after we die—which is a mon-strously heartless thing for it to do. We just can't imagine a world without "I."

How do we cope with this bleak truth? For three years, Wis-consin native Eric McClean posted videos on YouTube—108 in all—detailing his fight against leukemia. On August 14, 2012, he posted his final video, tearfully confessing that he had lost his battle and was entering hospice care. "I'm so scared. God, is it scary," he quavered, choking back sobs. "I'm gonna miss everyone, my family, my brothers and my sister, my parents, my wife, Cari. . . . I tried so hard, fought so hard. But the fight is over." He died nine days later at twenty-eight years old.

Nick, too, confronted the void of nonexistence, admitting to me during yet another hospital stay that, during his darkest days, he seriously considered suicide. In the end, he said, he chose to think of his approaching death as surfing one last giant wave: "I decided I'm just gonna ride that wave all the way into shore."

That, I think, is the key. Everyone finds their own way to create meaning out of our allotted time on this Earth—an evocative metaphor, a personal philosophy, a belief in an afterlife, anything that shapes the ups and downs of our lives and inevitable deaths into a story that makes sense.

We scattered Nick's ashes, half in New York and half in Los Angeles, at his request. He loved both cities for different reasons and had always dreamed of being bicoastal. And then? Life went on without him. The world kept turning. Yet while my friend is gone, he is not forgotten—not as long as there is someone who loved him to tell his story.

9

The Accidental Fabulist

She realized as a girl of eight that if she sat down and
wrote her stories, she could escape the parts of life
she didn't like, embroider the parts she did and thus
control the life she had.

—DUDLEY CLENDINEN, *A Place Called Canterbury*

I can't recall the exact moment I discovered the children's en-
cyclopedia set that graced the top of our hallway bookshelf,
but those white-bound volumes set my mind on fire and sparked
a lifelong love of reading. By age four, I compulsively read the
text on cereal boxes over breakfast. In kindergarten, during
playtime, the teacher would find me huddled under a table with
a book, oblivious to the shrieking mayhem around me. My desire to read often overrode basic social skills. Once, in grade
school, a friend came over. Halfway through her visit, I wan-
dered off, found a quiet corner, and settled in with a pile of
books. When my friend asked what I was doing, I handed her a
book—a tacit invitation to join me—and kept on reading. Hon-
estly, it's amazing I had any friends at all.

My behavior in high school wasn't much different: I spent all my free periods in the library, feasting on its literary riches. It was there I first encountered Charles Dickens and the classic trilogy *Kristin Lavransdatter*; became enamored of Arthurian legend and French revolutionary history; and snuck a peek at the uncensored version of Chaucer's "Wife of Bath" tale. But my tastes weren't exclusively highbrow: I also devoured the novels of Taylor Caldwell, James Michener, and John Jakes, along with countless murder mysteries, fairy tales, and horror anthologies. My parents initially were bemused by this behavior and later bothered by it, fearing that my self-imposed isolation would stunt my social skills. At one point, in desperation, my father removed the door to my bedroom. But you can't force a kid who loves books that much not to read. Trust me—we will find a way.

My parents need not have worried, if the latest cognitive neuroscience research is any indication. I was a late bloomer, socially speaking, but my compulsive love for reading may have conferred other advantages. In 2011, Canadian psychologist Raymond Mar found that the brain networks associated with stories overlapped significantly with the regions we use to navigate social interactions. He believes that this capacity helps us identify with the thoughts and feelings of others, making it easier to predict their intentions and actions. In fact, Mar and his cohort, Keith Oatley, found that people who read a great deal of fiction, in which omniscient narration is common, are better able to empathize with others and view a given situation from another's perspective than those who do not. Thank you, science, for validating my childhood bibliophilia.

Granted, fiction is not the only factor influencing whether we become empathetic creatures, any more than nature (or nurture) is the sole author of our behaviors and traits. The ability to comprehend another's viewpoint and project how that person might respond to a given situation also makes it much easier to manipulate other people for selfish ends. The point is that story-telling isn't just for entertainment: it's also a powerful social and psychological tool that helps us make sense of the world. Like all such tools, it can be used for good or evil.

"The function of the human mind is to make rough approximations of the world, telling stories that allow us to predict what other actors are going to do," Andrew Gerber, a psychologist at Columbia University Medical Center, explained when I visited his office on Manhattan's Upper West Side. One way we do this is via "schemas": sets of predictive models that enable us to make educated guesses as we navigate our social world, adapting our own behavior in response to how we expect others to behave. "Schemas are most often emergent, in the sense that you can't point to a single origin of the story," said Gerber. "Rather, it was the coming together of many different factors over a period of years that led to the formation of an effective story."

We instinctively sort people into broad categories based on past experience. The concept is similar to stereotyping, only less overtly prejudicial; it's more of a social survival mechanism. When we walk into a party, we scan the room and make quick assumptions about the other guests as we decide whom to approach and how to break the ice. We don't have time to carefully assess every last detail about every single guest, so we take shortcuts. If we have an in-depth conversation with a particular guest and gather more data—otherwise known as becoming better acquainted—we will tell them some of our stories, and

THE ACCIDENTAL FABULIST 263

they will tell us some of theirs, and we will integrate the new information into our schema for that person.

Caltech neuroscientist Christof Koch pointed out that these learned predictive models—which he calls "priors"—become more entrenched the longer we know someone, influencing our interpretation of events and our memories. Invariably, he said, we will remember details consistent with our existing priors because to do otherwise would disrupt the narrative we have constructed about someone's identity and behavior. "It would make for a far messier story," he explained. "We like clean stories, and we like happy endings."

The course of conversation does not always run smooth. We have all found ourselves in the awkward and frustrating position of trying and failing to connect with someone despite our best efforts. The feeling contrasts sharply with the pleasure we derive from a lively, free-flowing conversation with someone we perceive to be on our same wavelength. That sense of connection may be all in our heads. Princeton University neuroscientist Uri Hasson's specialty is exploring the dynamics of "interacting brains," performing fMRI scans of human subjects as they watch movies or listen to personal stories. He made headlines in 2010 with his experiments demonstrating "speaker-listener neural coupling"—a kind of Vulcan "mind meld" between the storyteller and the listener.

First Hasson had a graduate student, Lauren Silbert, tell an engaging story about her comically disastrous high school prom (featuring two suitors, a fist fight, and a car accident—quite the prom night) while inside an MRI machine. Her voice was recorded with a microphone rigged to filter out the machine's noise, and the taped story was then played back to eleven "listener" volunteers, also while in the MRI. All the listeners showed similar brain activity—they seemed to respond to the same elements in the story—but they also showed similar brain activity to Silbert (the storyteller). Hasson thinks there might be more overlap than previously believed between the brain's "production and comprehension" systems. It is possible that the brain processes complicated audio input like language through a dual mechanism that creates its own version of the signal and then compares it to what it "heard."

There was a bit of a time delay between storyteller and listener responses—just enough time to allow for the flow of information between two brains. That, said Hasson, indicates causality: to some extent, the storyteller's words shape the responses in the listener's brain. There was also a subset of brain regions that lit up for some listeners *before* the corresponding activity in the storyteller's brain—as if those listeners were actively predicting or anticipating the next part of the story. The strength of the effect might be related to how well people understand one another. Hasson has found that the stronger the coupling between the storyteller and the listener's brain response, the better the degree of understanding. So that feeling you get in conversation when the two of you just seem to "click"? That corresponds with brain activity: brain responses become more similar the more two people understand each other.

Nor does the brain distinguish much between reading about an experience, or listening to a description, and actually encountering it. An Emory University study scanned the brains of seven college students as they felt objects with different textures. Then the students listened to a sampling of phrases evoking the sense of touch—some very literal, others metaphorical—and found that while the language-processing regions showed activity for all the phrases, the metaphorical phrases also caused the region associated with touch to light up. This suggests an intriguing correlation between sensory perception and how we respond to metaphors. It may offer a clue as to why human beings love stories so much. A really good narrative engages our senses as well as our language-processing skills. The more brain regions are involved, the more vivid our experience will be—and the more likely we will be to recall the details.

This Is Your Life

My mother claims she always knew I would become a writer. At seven, I filled a notebook with early attempts at poetry, an effort that has mercifully been lost forever, along with most of my "juvenilia." At ten, I moved on to penning fables and horror stories, and at thirteen, I wrote a romance novel with a steamy love scene, even though the full sexual import was lost on me at that age. My mother naturally took me for a genius. She was mistaken. I may have been precocious, but deep down, we are all raconteurs, drawing on past memories and weaving them into a coherent narrative to construct our autobiographical selves.

"The story metaphor fits lives," said Dan McAdams, a psychologist at Northwestern University who specializes in the autobiographical self. "It has a beginning, middle, and end. It has characters with times and scenes. That is how life is, and that is how people see life." McAdams has identified three distinct layers to his model for the autobiographical self. By the age of two, most of us can recognize ourselves in mirrors and understand how we fit in with relation to others. At this point, we are Actors in our personal narratives, defining ourselves by explicit traits and the roles we play. We might be shy and conscientious students, while others are funny, outgoing, and popular.

Around the age of eight, we add another layer: the self as Agent. Now, in addition to being actors in our own lives, we also perceive our own agency: we can look at our past, project into the future, and set goals, whether we want to become a physicist, a writer, or merely find a best friend. Finally, as we move into early adulthood, we embrace the self as Author, developing a narrative identity that we continue to hone for the rest of our lives. "You create a narrative for your life that explains what kind of actor you are, and why, as an agent, you do what you do," said McAdams. "Autobiographical reasoning enables you to look at the stories in your life and see the underlying theme or motif."

His conclusions are based on hundreds of personal life-story interviews, conducted over many years, with adults from all walks of life. The entire process lasts two hours. The interviews are recorded and transcribed, and McAdams works from the written transcripts. The subjects are asked to imagine their lives as a book, with chapters, just like a novel. Then McAdams asks them to focus on key scenes: a high point, a low point, a turning point, a negative early memory, a positive early memory, and so on—all universal elements to a good narrative. "That is the best part of the interview because you get these really rich stories within the bigger story," he admitted.

I would probably select earning my black belt in jujitsu—the culmination of six years of intense training, weathering broken toes, a sprained wrist, a nasty head injury, and countless bruises—as a high point in my life, although it wasn't exactly pleasant. My *shodan* test spanned a grueling three hours. I injured my elbow two-thirds of the way through, but my instructors didn't stop the test. They just iced the elbow to control the swelling in between matches before tossing me back onto the mat to face my next opponent, smacking me occasionally with a bamboo stick when my energy flagged. Several friends found this difficult to watch, although my petite, pastel-clad, gray-haired mother shocked everyone by shouting and cheering me on with relish from the sidelines, confident that I could handle the blows. I could barely stand by the time the grand master tied the black belt around my waist, but the aches and pains were all forgotten as my cheering classmates hoisted me in the air triumphantly. Physically, it was one of the most difficult things I've ever done; that's what made the accomplishment so meaningful.

What about a low point? There was that humiliating moment at a high school piano recital when my mind went blank right in the middle of a Bach prelude and fugue and I had to stop without finishing the performance. But such moments, while cringe-worthy, pale in comparison to the death of my friend Nick, which forced me to confront the stark reality of human mortality. I would probably pick my decision to move to New York City after college and my marriage as major turning points; both choices significantly changed the course of my life in ways I could never have predicted, most definitely for the better.

A negative early memory might be the first time I told a lie, claiming I'd had a baloney sandwich for lunch instead of peanut butter because I knew it would please my mother, who was always urging me to broaden my finicky palate. I was soundly punished, even though, just shy of five, I didn't quite understand the concept of a lie. Clearly it was something very bad, and the lesson took: I am a terrible liar to this day. For a positive early memory, I'd choose visiting my grandmother's house in Maine, where I first tasted her wild rabbit stew and missed a family excursion to town for ice cream because I lost track of time exploring the surrounding woods. Someone else would have an entirely different set of examples—even I might choose different memories on a different day—but the broad strokes are the same: we all experience many such moments over the course of our lives, and we weave them into our story as it evolves.

Next, McAdams asks the subject to identify those people who have filled the roles of heroes and villains. That high school biology teacher who encouraged my scientific curiosity, the swim coach who pushed me to improve my backstroke, and the English professor whose passion made *Beowulf* and Milton come alive in the classroom would all qualify as heroes. Villains might include that girl in junior high who scrawled nasty, untrue things about me on the bathroom wall (and borrowed my pen to do it), or the ex-boyfriend who went home early after our Valentine's Day dinner because he was "tired" and spent the night with another woman instead. Subjects are also asked to think about future chapters—goals and aspirations—and how their values and beliefs are reflected in the full arc of the personal narrative.

Finally McAdams asks subjects to identify overarching themes running through their stories. One common theme is redemption, particularly among people he calls "highly generative"—those who volunteer in soup kitchens or political campaigns, start their own nonprofit charity, or otherwise seek to have a positive effect on the world. Their stories invariably involve hardship and suffering, but with an optimistic twist: they triumphed over their woes, learned valuable life lessons from the pain, and emerged stronger for it. This doesn't mean highly generative people are necessarily better storytellers: "People who are lower in generativity tell very interesting, dramatic stories, but they are not as redemptive," McAdams said. Being a generative person requires a lot of hard work—"It would be so much easier to just stay home and watch *American Idol*"—and he thinks that having a strong redemptive narrative serves as a motivational tool.

My highly generative mother is a case in point. She lost her father to a bad batch of bathtub gin at a very young age. Her mother didn't have the means to raise all eight children alone, so the siblings were farmed out to a succession of foster homes, and my mother ended up in a Catholic orphanage in Maine. She spent much of her childhood in abject poverty. Canned tomatoes on toast were a special treat, and she and her sisters used to pull pieces of tar off the roof to substitute for chewing gum. Many people who hear my mother's stories express surprise that she isn't bitter, but to this day she insists that her time at the orphanage gave her some of the happiest memories of her life— even if she did get kicked out of the choir for being hopelessly tone deaf. As an evangelical Christian, my mother also has a literal redemption story: her conversion and newfound faith are what sustained her after losing her best friend to the recklessness of a drunk driver. All these elements helped shape her personal narrative and influenced the person she has become.

While themes may vary from person to person, all stories are ultimately about change. "Nobody tells a story about continuity, about how they've always been the same," McAdams said. We can change our stories, thereby changing ourselves in some small but significant way, even though our core self remains the same. For Andrew Gerber, the core self is comprised of preexisting constraints—the nature part of the equation, in which genes and synapses impose limits, much like the rules that define whether a poem is a haiku, a sonnet, or a villanelle. "Self is the content of what one builds out of those constraints," he said. We have a great deal of leeway within those constraints to tell different stories. The personal narrative I have constructed has most certainly changed over the years. The facts might be the same, but my interpretation of them has altered because I have more contextual information, enabling me to see new connections and affording me a broader perspective.

Psychological change is gradual and incremental, often requiring months or years of therapy. Gerber has found that when he asks patients to pinpoint a specific moment of change, they can't answer right away. When pressed, they offer a specific memory that usually involves the clinician breaking out of his or her professional role, sharing a personal detail, perhaps. But that choice is rather arbitrary. "Because we don't have a good model for truly gradual changes, we attach our sense of change onto a moment," he explained. "And that moment comes, in our memory system, to represent a turning point."

The Persistence of Memory

When I was young, my family spent the day at Green River Gorge in Washington State, an area popular with white water river rafters. I was keen to taste the mountain water, and my brother—a fledgling Boy Scout, quite proud of his new outdoor knowledge—informed me that the water was coldest and freshest where the current was strongest. So I marched to just such a spot and leaned over to dip my hand in the water. My brother had not warned me about the slippery bank, however, and I toppled into the river, the powerful current carrying me downstream as I madly dog-paddled to keep my head above the water.

My father jumped in to rescue me and I figured I was safe. Such is the innocent perspective of a child. But my mother, standing on the bank, could see the fear on my father's face and realized the gravity of the situation: he couldn't fight that current either. Fortunately there were many people at the gorge that day, and they formed a human chain to pull us to safety. I bounced back quickly from the experience, and it is now an amusing childhood story. But my mother has never been back to Green River Gorge. To her, it is a painful reminder that she almost lost her husband and child.

My mother and I often recall the same events differently, because we experienced those events from different perspectives. If a geologist, an ornithologist, and a botanist go for a walk in the park, when they return, they will likely give very different accounts of this shared experience. The geologist might not have noticed the birds, while the ornithologist might not have noticed the plants and the botanist might not have noticed unusual rock or sediment. The British psychologist Sir Frederick Bartlett used this scenario back in the 1930s to demonstrate how different perspectives or points of view inevitably color one's recollection of an event. It is a familiar occurrence for police officers accustomed to interviewing witnesses. If there are four eyewitnesses to a car accident, you will get four separate accounts of how it happened. The basic facts might be the same, but each witness will zero in on different details.

I would wager that my mother's memory of that day in Green River Gorge is far more vivid than mine too, because her emotional fear response was so much greater. Powerful emotions make for stronger connections and associations, and for more vivid memories. Blame the amygdala and its powerful neurochemicals (dopamine, serotonin, epinephrine, norepinephrine, and acetylcholine), which serve as a primitive alarm system alerting the brain to potential danger or stress—the so-called fight-or-flight response.

The cells that produce those chemicals are found in the brain stem, from which axons branch out into every other area of the brain to produce a coordinated sensory response. Fear and anxiety are the result of those cells flooding the neurons in various brain regions with neurochemicals, although only active cells are affected. Different regions process and store different sensory aspects of a given experience, and this widespread flooding of neurochemicals ensures that all those regions record information from the event. The cortex then integrates the information, and the experience becomes part of our memory. Those memories, in turn, may influence how we react in the future—such as my mother's refusal to return to Green River Gorge.

But how accurate are those memories? Like many people, I have vivid recollections of the devastating 2001 terrorist attacks on New York City and Washington, D.C. There were the initial shock and horror; anxiety and dread over the safety of friends; relief for those who survived, and grief for those who did not. There was the eerie quiet that descended on Washington, D.C., as the entire city shut down, broken only by the sound of military helicopters flying overhead. Then there were the memories of the aftermath: exhausted, emotionally numbed friends pulling multiple double shifts in the hospitals to sift through the gooey mess of mangled body parts in hopes of finding some clue to identification; the thick cloud of smoke that hung over New York City; the stench of burnt, decaying flesh that wafted from Manhattan into the outer boroughs for months after the tragedy. And of course, there was funeral after funeral after funeral.

Psychologists term these recollections of emotionally charged, significant events "flashbulb memories" because we feel as though we have taken a snapshot of those moments, incorporating both the factual details and our emotions at the time. Days after the 9/11 attacks, two psychologists in New York decided to survey a few thousand Americans across seven different cities, asking them to report everything they could remember about that day. Then they followed up with identical surveys eleven months and three years later to see how long those memories lasted. The accuracy of the participants' memories, compared to their initial accounts right after the attacks, dropped to 60 percent at the eleven-month mark and 50 percent after three years. In general, they recalled details of time and place more accurately (80 percent), while their memories of their emotions at the time were only 40 percent accurate. They also believed that they recalled that day far better than they actually did, and the vividness of those memories—whether accurate or not—remained undiminished.

Just because a memory is vivid, that doesn't mean it's accurate. The same is true for how well we remember a fictional story. Sir Frederick Bartlett conducted a classic experiment, in which he told participants a Native American folktale called "The War of the Ghost," and later asked them to retell it. He found they would remember the tale accurately the first time they retold it, but with each retelling, that accuracy diminished. The story became shorter or the subjects forgot it. They dispensed with the supernatural aspects in favor of a more conventional English fairy tale; a few even tacked on a moral at the end, as if it were a fable.

Bartlett was the first to realize that remembering a fictional story is very similar to how we remember an event from our past and construct a story about that. WUSTL's Henry Roediger, a psychologist who studies memory, explained that memory does not operate like a computer, where stored files are pulled and reopened intact. "It is much more like storytelling, where you are given a few cues and use those to make up the story," he said. "It is probably mostly right, consistent with your past knowledge, but you can add elements with each retelling." Our memories are distributed over several different regions of the brain, and each time we recollect an event, we are, in essence, reconstructing it from scratch, based on a few key clues. Some of the details that emerge over time might be pure fabrication, but we still believe they are accurate.

In fact, neuroscience demonstrates that as far as our brains are concerned, imagining the future is the same as recalling the past. Amnesic patients who have lost their autobiographical memories also lose their ability to imagine the future. Roediger's wife, Kathleen McDermott, a fellow WUSTL psychologist, scanned the brains of subjects as they either recalled a past event or imagined a future one, usually a very common experience, such as a birthday party or getting lost. To her surprise, the recorded brain activity produced strikingly similar patterns.

So the brain regions involved in autobiographical memory are also involved in "episodic future thought," with significant implications for the accuracy of our memories—and, by extension, for the truthfulness of our personal narratives. If these two activities are so similar in brain activity patterns, what keeps us from imagining that we remember something that never happened?

Inception Is Easy

It was the summer of 2010 when the dystopian science-fiction film *Inception* ruled the box office, despite its dark themes and convoluted plot involving a special band of thieves who break into people's subconscious minds to steal information as they dream. My husband, Sean, and I loved the film and shared our enthusiasm with psychologist Carol Tavris over dinner one night. We especially savored the careful attention to physics details, notably a scene in an elevator that served as the perfect cinematic depiction of Einstein's equivalence principle. But for Tavris, the very premise—that it would be supremely difficult to implant an original idea in other people's minds such that they believe it is their own (the "inception" of the title)—was ludicrous, making it impossible for her to suspend her disbelief. "Inception is easy," she declared.

It is the issue of authorship that is significant. As the film makes clear, planting an idea in someone's head is simple enough: tell people not to think of pink elephants, and chances are that images of pink elephants will spring to mind. But they know that the pink elephants came from an outside suggestion. Tavris's point was that it is just as easy to manipulate people into thinking the pink elephants were their idea all along.

Elizabeth Loftus, a psychologist at the University of Washington, has spent much of her career studying memories, particularly false ones. She was at the forefront of debunking so-called repressed memories in the 1980s and 1990s, when numerous high-profile cases emerged of women in therapy who had supposedly recovered memories of being sexually abused as children. After a therapist employed hypnosis and staged mock exorcisms, a Wisconsin nurse named Nadean Cool became convinced that she had belonged to a Satanic cult as a child, had been raped, and even witnessed the brutal murder of her eight-year-old friend. None of it was true, and when Cool realized what had happened, she sued her therapist and won $2.4 million in damages in 1997. It is an extreme case, but not an isolated one.

Friends, family, and therapists can alter our recollection of events, even if they do so unintentionally, because we look to others to corroborate details. One of Loftus's experiments required subjects to read four short stories about events in their lives, each recounted by an immediate relative—all true except for one fictional event, an account of being lost in a shopping mall. None of the subjects had ever been lost in the mall as children, but they were told that close relatives had confirmed all four stories. There were four parts to this manufactured "memory": being lost for an extended period, breaking down in tears, being comforted and helped by an elderly woman, and eventually being reunited with one's family.

The key to implanting a false memory is to start with an element of truth—in this case, an actual shopping trip that had taken place, although the subject had not gotten lost. Because the subject knew the family member had provided the stories, there was corroboration by another person. People also tend to forget the source of misinformation after they incorporate it and mistake it for part of the original experience. Finally, asking someone to imagine a childhood event—a common therapeutic technique used to "recover" lost memories—has been shown to induce false memories. Imagining such an event makes it seem more real to the patient and increases their confidence that it actually occurred. Nearly one-third of participants in the Loftus study claimed to recall being lost in the mall as children, and 25 percent stubbornly clung to that belief, insisting it had really happened, even when they were told that it was false. A false memory had been implanted merely by the power of suggestion, and that memory seemed as real as events that actually happened.

See? Inception is easy. Blame your storytelling brain. When confronted with details that don't seem to fit, the brain will create a narrative in which everything makes sense, regardless of whether it is true. Neuroscientist Michael Gazzaniga arrived at this conclusion through his work with so-called split-brain patients—those who had undergone surgery to sever the corpus collosum, the fabric that connects the right and left hemispheres of the brain. Patient Zero was a veteran of World War II dubbed W. J. who suffered from severe epilepsy and opted to have the surgery.

Gazzaniga tested W. J. before and after the surgery with a tachioscope, a device that presents visual stimuli to one eye but not the other, in order to establish whether the two hemispheres of the brain are communicating properly. W. J. could identify objects in either hemisphere prior to the surgery, but afterward, he was unable to describe objects presented only in his left visual field, even though both of his eyes functioned perfectly. The sensory input simply did not register.

Then Gazzaniga showed W. J. two different pictures—a chicken claw to the left and a snow-covered scene to the right—and presented a set of images, asking him to choose the one most closely associated with the prior image he'd seen. W. J. pointed at two different pictures: one hand pointed to a picture of a chicken, while the other hand pointed to a shovel.

The choice of the chicken was unsurprising, since the left hemisphere had "seen" the chicken claw, but it didn't explain the choice of the shovel. When asked why he also chose the shovel—the snow-covered scene had only been "seen" by the right hemisphere—W. J. announced that one would need it to clean out the chicken coop. His mind created a story to incorporate the detail that didn't seem to fit.

Gazzaniga calls the left hemisphere the brain's interpreter: it is responsible for creating a plausible narrative to eliminate any cognitive dissonance or discrepancies between the many different sensory inputs. "It is the left hemisphere that engages in the human tendency to find order in the chaos, that tries to fit everything into a story and put it into a context," he explained in his book *Who's in Charge?* "It seems driven to hypothesize about the structure of the world even in the face of evidence that no pattern exists. Facts are great, but not necessary." It is yet another reason why our memories, and the personal stories we weave from them, are not as factual as we believe them to be.

Myth Makers

In January 2012, NPR's *This American Life* aired an excerpt from Mike Daisey's one-man stage show *The Agony and the Ecstasy of Steve Jobs* a moving account of the appalling conditions faced by Chinese workers in Apple's overseas factories—the all-too-human cost of our iPhones and iPads. That excerpt, which quickly became one of the most popular pieces on NPR's Web site, included the heartbreaking account of an old Chinese man with a ruined hand—an injury sustained on the job—touching an iPad for the very first time and marveling, "It's a kind of magic."

There was just one problem: the story wasn't true. Neither were many other details, as an intrepid reporter with American Public Media discovered when he tracked down Daisey's interpreter, Cathy Lee. She flatly contradicted many of Daisey's statements. Eventually Daisey broke down and admitted—in a painful on-air confrontation with host Ira Glass—that he had fudged some of the facts in pursuit of a more powerful performance piece. NPR retracted the excerpt, and Daisey expressed regret for allowing the performance to be passed off as factual reporting, although he later wrote on his blog that, despite the controversy, he stood by his work: "My show is a theatrical piece whose goal is to create a human connection between our gorgeous devices and the brutal circumstances from which they emerge." Had he labeled it as such from the outset, there may not have been a controversy.

Even though Daisey copped to changing certain details, he doubled down on several remaining discrepancies between his account and Lee's—notably, his exchange with supposedly underage workers. There have been documented cases of underage workers in Apple's Chinese factories, but Lee insisted that Daisey never interviewed any. Daisey claimed he had. Who should we believe? Psychologist Nate Kornell has compared storytelling to the game of telephone many of us played as children. "The second time I tell a story, what I'm remembering is the first time I told the story," he wrote in *Psychology Today*. "And the 201st time, I'm really remembering the 200th time. Many of our memories are records of our own stories, not of events that actually took place."

Daisey told his story hundreds of times, night after night, polishing and embellishing the details until he could no longer distinguish between fact and fiction. That, wrote Kornell, is a recipe for a false memory. "Even if it never happened, it would make it completely real to him." Since she did not tell and retell the same story many times, it is Lee who would give the more plausible account. This is why eyewitness testimony is often deemed unreliable after the first interview: after a witness has told a story once, the account becomes increasingly tainted with each subsequent retelling.

Cases like Daisey's are why Jonathan Gottschall, a professor of English and author of *The Storytelling Animal*, insisted to *Scientific American* that all memoirs should come with a disclaimer: "Based on a true story." In many cases, Gottschall claimed, our life stories are "boldly fictionalized . . . based on distorted memories and wildly optimistic assessments of our own qualities." No personal narrative will ever be a purely objective account. The personal is inherently subjective, and we will often choose to tell the version of events that is most flattering to ourselves. We fabricate and embellish even when we believe ourselves to be truthful. We are accidental fabulists.

E. M. Forster once observed, "Fiction is truer than history because it goes beyond the evidence, and each of us knows from our own experience that there is something beyond the evidence." Therein lies the value of the autobiographical self. It is less of a bug and more of a feature when it comes to constructing a unified whole out of the many components that contribute to our sense of self. You can sequence my DNA, scan my brain, map my connectome, subject me to any number of psychological tests, even hypothetically map out the position and velocity of every single subatomic particle that makes up my mind and body, but you won't find my essence in any one of them alone. If you took a digital image of me and broke it down into the component pixels, in one sense, you would have a complete description of that image, neatly rendered in zeroes and ones. But that information is meaningless without an interpretive layer that enables you to see the complete picture. Stories provide that unifying interpretive layer.

Our personal narratives have more in common with myth than with straight reportage: stories that may not be accurate in every detail but embody an ideology or greater truth. An unjustly ignored novel by C. S. Lewis called *Till We Have Faces* retells the Cupid and Psyche myth from the perspective of Psyche's ugly elder sister, Orual, who eventually becomes queen of Lewis's fictional realm.

Despite her role in bringing about her sister Psyche's downfall in this retelling, Orual is a good queen and a sympathetic character. But the book ends with a shattering moment of self-awareness, when the dying Orual—who has long held a grudge against the gods for their perceived unjust treatment—finally has the opportunity to appear before them and read her "complaint," a book she carefully composed over the course of her life. It is the mythology she has created of her experience, the story she tells about herself to the world. Yet in the presence of the eternal, she realizes that her once-great work is nothing more than "a little, shabby, crumpled" parchment, filled not with her usual elegant handwriting, but with "a vile scribble— each stroke mean and yet savage." This is her true voice, her true self, stripped of all the delusions and lies she has been hiding behind all those years. "I saw well why the gods do not speak to us openly, nor let us answer," Orual concludes. "Till that word can be dug out of us, why should they hear the babble that we think we mean? How can they meet us face to face, till we have faces?"

Lewis was a mythmaker. He understood the innate human

need to tell stories. He also knew that human beings are fallible and prone to self-delusion. Like Orual, we spend our lives weaving a story about ourselves. It is worked and reworked constantly. True self-knowledge come when we finally manage to hit the truth of the matter and say what we really mean.

Do you really want to know who I am?

Let me tell you a story.

BIBLIOGRAPHY

Prologue

Galton, Francis. *English Men of Science: Their Nature and Nurture*. London: Macmillan and Co., 1874.

Harris, Judith Rich. *No Two Alike: Human Nature and Human Individuality*. New York: W. W. Norton, 2006.

———. *The Nurture Assumption: Why Children Turn Out the Way They Do*. New York: The Free Press, 1998.

James, William. *Psychology: The Briefer Course*. Mineola, NY: Dover Publications, 2001.

Keller, Ellen Fox. *The Mirage of a Space Between Nature and Nurture*. Durham, NC: Duke University Press, 2010.

LeDoux, Joseph. *The Synaptic Self: How Our Brains Become Who We Are*. New York: Penguin, 2003.

Lewontin, Richard. "It's Even Less in Your Genes," *The New York Review of Books*, May 26, 2011.

de Montaigne, Michel. *The Complete Essays*. New York: Penguin Classics, 1993.

Nettle, Daniel. *Personality: What Makes You the Way You Are*. Oxford, UK: Oxford University Press, 2007.

Ridley, Matt. *Nature via Nurture: Genes, Experience, and What Makes Us Human*. New York: Harper Collins, 2003.

I. ME

1. What's Bred in the Bone

Aldhous, Peter. "U.S. Court Ruling on Breast Cancer Genes a Mixed Blessing," *New Scientist*, June 13, 2013.

Angrist, Misha. *Here Is a Human Being: At the Dawn of Personal Genomics*. New York: Harper Collins, 2010.

Bufe, B. et al. (2005) "The Molecular Basis of Individual Differences in Phenylthiocarbamide and Propylthiouracil Bitterness Perception," *Current Biology* 15: 322–27.

Callaway, Ewen. "Soapy Taste of Coriander Linked to Genetic Variants," Nature.com, September 12, 2012.

Chouard, T. (2010) "Evolution: Revenge of the Hopeful Monster," *Nature* 463: 864–57.

Cohen, Andrew. "Nature vs. Nurture: The Continuing Saga of the Gene Patenting Case," *The Atlantic*, April 2004.

Dahl, Melissa. "Your Cilantro Love or Hate May Be Genetic," *Body Odd*, MSNBC.com, October 6, 2011.

Dahm, R. (2008) "Discovering DNA: Friedrich Miescher and the Early Years of Nucleic Acid Research," *Human Genetics* 122: 565–81.

———. (2005) "Friedrich Miescher and the Discovery of DNA," *Developmental Biology* 278(2): 274–88.

Do, C. B. et al. (2011) "Web-Based Genome-Wide Association Study Identifies Two Novel Loci and a Substantial Genetic Component for Parkinson's Disease," *PLoS Genetics* 7(6): e1002141.

Drayna, D. (2005) "Human Taste Genetics," *Annual Review of Genomics and Human Genetics* 6: 217–35.

Duffy, V. B. et al. (2004) "Bitter Receptor Gene (TAS2R38), 6-n-propylthiouracil (PROP) Bitterness and Alcohol Intake," *Alcoholism: Clinical and Experimental Research* 28: 1,629–37.

Eiberg, H. et al. (2008) "Blue Eye Color in Humans May Be Caused by a Perfectly Associated Founder Mutation in a Regulatory Element Located Within the HERC2 Gene Inhibiting OCA2 Expression," *Human Genetics* 123: 177–87.

Eriksson, N. et al. (2012) "A Genetic Variant Near Olfactory Receptor Genes Influences Cilantro Preference," http://arxiv.org/abs/1209.2096 (Preprint).

Flam, Faye. "Why We Hate Broccoli," Planet of the Apes, *The Philadelphia Inquirer*, December 19, 2011.

Goetz, Thomas. "23andMe Will Decode Your DNA for $1000. Welcome to the Age of Genomics," *Wired*, November 17, 2007.

Goldschmidt, R. (1933) "Some Aspects of Evolution," *Science* 78: 539–47.

Hayes, J. et al. (2011) "Allelic Variation in TAS2R Bitter Receptor Genes Associates with Variation in Sensations from and Ingestive Behaviors Toward Common Bitter Beverages in Adults," *Chemical Senses* 36: 311–19.

Henig, Robin Marantz. *The Monk in the Garden: The Lost and Found Genius of Gregor Mendel, the Father of Genetics*. New York: Houghton Mifflin/Mariner Books, 2001.

Herper, Matthew. "The Future Is Now: 23andMe Now Offers All Your Genes for $999," *Forbes*, September 27, 2011.

Hollingworth, P. et al. (2011) "Common Variants at ABCA7, MS4A6A/MS4A4E, EPHA1, CD33 and CD2AP Are Associated with Alzheimer's Disease," *Nature Genetics* 43: 429–35.

Judson, Olivia. "The Monster Is Back, and It's Hopeful," *Opinionator, The New York Times*, January 22, 2008.

Katsnelson, Alla. "Genetics Tells Tall Tales," Nature.com, July 10, 2010.

Kayser, M. et al. (2008) "Three Genome-Wide Association Studies and a Linkage Analysis Identify HERC2 as a Human Iris Color Gene," *American Journal of Human Genetics* 82: 411–23.

Keim, Brandon. "Genome Structure," *Wired*, July 29, 2011.

———. "The Scientific Search for the Essence of a Tasty Tomato," *Wired Science, Wired*, May 16, 2012.

Kim, J. et al. (2009) "The Role of Apolipoprotein E in Alzheimer's Disease," *Neuron* 63: 287–303.

Kim, U. et al. (2003) "Positional Cloning of the Human Quantitative Trait Locus Underlying Taste Sensitivity to Phenylthiocarbamide," *Science* 299(5610): 1221–225.

Knaapila, A. et al. (2012) "Genetic Analysis of Chemosensory Traits in Human Twins," *Chemical Senses* 37: 869–81.

Lai, Chao-Qiang. "How Much of Human Height Is Genetic and How Much Is Due to Nutrition?" *Scientific American*, December 11, 2006.

Lewis, Ricki. "Why I Don't Want to Know My Genome Sequence," *DNA Science, PLoS Blogs*, November 1, 2012.

Marshall, Jessica. "The Color Code," *New Scientist*, March 10, 2007.

Mauer, L., and A. El-Soheny. (2012) "Prevalence of Cilantro (*Coriandrum sativum*) Disliking Among Different Ethnocultural Groups," *Flavour* 1(8).

McClelland, Susan. "I Will Die the Most Horrible Death," *The Guardian*, August 10, 2009.

McGee, Harold. "Cilantro Haters, It's Not Your Fault," The Curious Cook, *New York Times*, April 4, 2010.

McInerney, N. M. (2010) "Evaluation of Variants in the CHEK2, BRIP1 and PALB2 Genes in an Irish Breast Cancer Cohort," *Breast Cancer Research Treatment* 121: 203–10.

Meijers-Heijboer, H. (2002) "Low-Penetrance Susceptibility to Breast Cancer Due to CHEK2(*)1100delC in Noncarriers of BRCA1 or BRCA2 Mutations," *Nature Genetics* 31: 55–9.

Miura, K. et al. (2007) "A Strong Association Between Human Earwax-Type and Apocrine Colostrum Secretion from the Mammary Gland," *Human Genetics* 121: 631–33.

Myers, P. Z. "It's More than Genes, It's Networks and Systems," *Pharyngula*, July 24, 2010.

———. "Steve Pinker's Hair and the Muscles of Worms," *Pharyngula*, December 19, 2011.

Naj, A. C. et al. (2011) "Common Variants at MS4A4/MS4A6E, CD2AP, CD33 and EPHA1 Are Associated with Late-Onset Alzheimer's Disease," *Nature Genetics* 43: 436–41.

Nierenberg, Cari. "Who Hates Cilantro? Study Aims to Find Out," *Body Odd*, MSNBC.com, May 16, 2012.

Ogilvie, Jessica Pauline. "Do It Yourself Genetics Tests," *Los Angeles Times*, April 18, 2011.

Onsekiz, C. (2001) "The Chemistry of Fresh Tomato Flavor," *Turkish Journal of Agricultural Forestry* 25: 149–55.

Ouellette, Jennifer. "Taster's Choice: Why I Hate Raw Tomatoes and You Don't," *Cocktail Party Physics, Scientific American*, May 29, 2012.

Pray, L. A. (2008) "Discovery of DNA Structure and Function: Watson and Crick," *Nature Education* 1.

Prodi, D. A. et al. (2004) "Bitter Taste Study in a Sardinian Genetic Isolate Supports the Association of Phenylthiocarbamide Sensitivity to the TAS2R38 Bitter Receptor Gene," *Chemical Senses* 29: 397–702.

Quynh, C. T. T. et al. (2010) "Influence of the Isolation Procedure on Coriander Leaf Volatiles with Some Correlation to the Enzymatic Activity," *Journal of Agricultural Food Chemistry* 58: 1093–99.

Reardon, Sara. "Court Ruling on Genes Is a 'Victory for Common Sense,'" *New Scientist*, June 18, 2013.

Roses, A. D. et al. (2010) "A TOMM40 Variable-Length Polymorphism Predicts the Age of Late-Onset Alzheimer's Disease," *Pharmacogenomics Journal* 10: 375–84.

Rotman, David. "DNA Sequencing to Go," *Technology Review*, February 17, 2012.

Rubenstein, Sarah. "Across the Land, People Are Fuming Over an Herb (No, Not That One)," *Wall Street Journal*, February 13, 2009.

Sturm, R. A., and T. N. Frudakis. (2004) "Eye Colour: Portals into Pigmentation Genes and Ancestry," *Trends in Genetics* 20: 327–32.

Sturm, R. A. et al. (2008) "A Single SNP in an Evolutionary Conserved Region within Intron 86 of the HERC2 Gene Determines Human Blue-Brown Eye Color," *American Journal of Human Genetics* 82: 424–31.

Sulem, P. et al. (2007) "Genetic Determinants of Hair, Eye and Skin Pigmentation in Europeans," *Nature Genetics* 39: 1443–452.

Tang, M. X. et al. (1998) "The APOE-Epsilon4 Allele and the Risk of Alzheimer Disease Among African Americans, Whites, and Hispanics," *Journal of American Medical Association* 279: 751–55.

Tepper, B. J. (1998) "6-n-propylthiouracil: a Genetic Marker for Taste, with Implications for Food Preference and Dietary Habits," *American Journal of Human Genetics* 63: 1271–276.

Thin Man, The. (1934) Metro-Goldwyn-Mayer. Dir. W. S. van Dyke. Screenplay by Albert Hacket and Francis Goodrich.

Tieman, D. et al. (2012) "The Chemical Interactions Underlying Tomato Flavor Preferences," *Current Biology* 22: 1035–39.

Wakamatsu, K. et al. (2008) "Characterization of Melanin in Human Iridal and Choroidal Melanocytes from Eyes with Various Colored Irides," *Pigment Cell & Melanoma Research* 21: 97–105.

Weischer, M. (2008) "CHEK2*1100delC Genotyping for Clinical Assessment of Breast Cancer Risk: Meta-Analyses of 26,000 Patient Cases and 27,000 Controls," *Journal of Clinical Oncology* 26: 542–48.

———. (2007) "Increased Risk of Breast Cancer Associated with CHEK2* 1100delC," *Journal of Clinical Oncology* 25: 57–63.

Wellcome Trust Case Control Consortium (2007) "Genome-Wide Association Study of 14,000 Cases of Seven Common Diseases and 3,000 Shared Controls," *Nature* 447: 661–78.

Wong, Kate. "Iceman's Genome Furnishes Clues to His Ailments and Ancestry," Observations, *Scientific American*, February 2, 2012.

Wooding, S. (2006) "Phenylthiocarbamide: A 75-year Adventure in Genetics and Natural Selection," *Genetics* 172: 2015–23.

X-Men: First Class. (2011) 20th Century Fox. Dir. Matthew Vaughan. Screenplay by Ashley Miller and Zack Stentz.

Yong, Ed. "How I Got My Genes Tested and the Birth of Science Writer Disease Risk Top Trumps," *Not Exactly Rocket Science*, *Discover*, July 21, 2010.

Yoshiura, K. et al. (2006) "A SNP in the ABCC11 Gene is the Determinant of Human Earwax Type," *Nature Genetics* 38: 324–30.

Zhang, S. (2008) "Frequency of the CHEK2 1100delC Mutation Among Women with Breast Cancer: An International Study," *Cancer Research* 68: 2154–157.

Zivkovic, Bora. "How Genotype Affects Phenotype," A Blog Around the Clock, *Scientific American*, September 17, 2011.

Online Resources

American Cancer Society: http://www.cancer.org/Research/CancerFacts Figures/index

Mendel's genetics: http://anthro.palomar.edu/mendel/mendel_1.htm

Mendel's genetics: http://naturalselection.0catch.com/Files/gregormendel .html

Supreme Court decision on gene patenting: http://www.supremecourt.gov/ opinions/12pdf/12-398_1b7d.pdf.

2. Uncharted Territory

Bardin, Jon. "Neuroscience: Making Connections," Nature.com, March 21, 2012.

Barres, B. (2010) "Neuro Nonsense," *PLoS Biology* 8: e1001005.

Bartels, A., and S. Zeki. (2004) "Functional Brain Mapping During Free Viewing of Natural Scenes," *Human Brain Mapping* 21: 75–85.

Bell, Vaughan. "A Guide to Neuroscience," *The Observer, The Guardian*, April 29, 2012.

———. "The Trouble with Brain Scans," *The Observer, The Guardian*, May 26, 2012.

Bennett, C. M. et al. (2010) "Neural Correlates of Interspecies Perspective Taking in the Post-Mortem Atlantic Salmon: An Argument for Multiple Comparisons Correction," *Journal of Serendipitous and Unexpected Results* 1: 1–5

Bilger, Burkhard. "The Possibilian," *The New Yorker*, April 25, 2011.

Blakeslee, Sandra, and Matthew Blakeslee. *The Body Has a Mind of Its Own: How Body Maps in Your Brain Help You Do (Almost) Everything Better.* New York: Random House, 2007.

Broca, Paul. (1863) "Localisations des Fonctions Cérébrales. Siège de la Faculté du Langage Articulé," *Bulletin de la Société d'Anthropologie* 4: 200–08.

———. (1861) "Sur le Principe des Localisations Cérébrales," *Bulletin de la Société d'Anthropologie* 2: 190–204.

Brookshire, Bethany. "Does Neuroscience Need a Newton?" *The Scicurious Brain, Scientific American*, December 3, 2012.

Burnett, Dan. "Neuroscience Fiction in Newspapers," *The Guardian*, May 1, 2012.

Bushwick, Sophie. "How Exactly Do Neurons Pass Signals Through Your Nervous System?" io9.com, January 19, 2012.

Cajal, Ramon y. *History of the Nervous System of Man and Vertebrates* (trans. Swanson, N., and L. W. Swanson). Oxford, UK: Oxford University Press, 1995.

———. *Recollections of My Life.* Cambridge, MA: MIT Press, 1989.

———. "The Structure and Connexions of Neurons," *Nobel Lectures, Physiology or Medicine 1901–1921.* Amsterdam: Elsevier, 1967.

Choi, Charles Q., and Txchnologist. "Brain Researchers Can Detect Who We Are Thinking About," *Scientific American*, March 14, 2013.

Cook, Gareth. "A Neuroscientist's Quest to Reverse Engineer the Human Brain," *Scientific American*, March 20, 2012.

Costandi, Mo. "The Discovery of the Neuron," *Neurophilosophy,* July 3, 2007.

———. "The Incredible Case of Phineas Gage," *Neurophilosophy,* July 6, 2007.

———. "Old Brains," *Neurophilosophy,* July 16, 2007.

———. "Phineas Gage's Connectome," *Neurophilosophy; The Guardian*, May 16, 2012.

———. "Phrenological Analysis of Ned Kelly's Death Mask," *Neurophilosophy,* August 19, 2008.

———. "Visual Images Reconstructed from Brain Activity," *Neurophilosophy,* December 12, 2008.

———. "Wilder Penfield, Neural Cartographer," *Neurophilosophy,* August 27, 2008.

Cowie, S. E. (2000) "A Place in History: Paul Broca and Cerebral Localization," *Journal of Investigative Surgery* 13: 297–8.

Damasio, Antonio. *Self Comes to Mind: Constructing the Conscious Brain*. New York: Pantheon Books, 2011.

Dronkers, N. F. et al. (2007) "Paul Broca's Historic Cases: High Resolution MR Imaging of the Brains of Leborgne and Lelong," *Brain* 130: 1432–441.

Eagleman, David. *Incognito: The Secret Lives of the Brain*. New York: Pantheon Books, 2011.

Elert, Emily. "The Man Who Wasn't There," *Discover,* October 2012.

Eliasmith, C. et al. (2012) "A Large-Scale Model of the Functioning Brain," *Science* 338: 1202–205.

Feinberg, Todd E. *Altered Egos: How the Brain Creates the Self*. Oxford, UK: Oxford University Press, 2001.

Feindel, W. (2007) "The Physiologist and the Neurosurgeon: The Enduring Influence of Charles Sherrington on the Career of Wilder Penfield," *Brain* 130: 2758–765.

Gill, A. et al. (2007) "Wilder Penfield, Pio Del Rio-Hortega, and the Discovery of Oligodendroglia," *Neurosurgery* 60: 940–48.

Gonzalez-Castillo, J. et al. (2012) "Whole-Brain, Time-Locked Activation with Simple Tasks Revealed Using Massive Averaging and Model-Free Analysis," *Proceedings of the National Academy of Sciences*, 109:5487–5492.

Greene, J. D., and J. M. Paxton. (2009) "Patterns of Neural Activity Associated with Honest and Dishonest Moral Decisions," *Proceedings of the National Academy of Sciences* 106: 12506–2511.

Gusnard, D. (2001) "Medial Prefrontal Cortex and Self-Referential Mental Activity: Relation to a Default Mode of Brain Function," *Proceedings of the National Academy of Sciences* 98: 4259–264.

Haynes, J. D. et al. (2007) "Reading Hidden Intentions in the Human Brain," *Current Biology* 17: 323–28.

Heatheron, Todd. (2011) "The Neuroscience of Self and Self-Recognition," *Annual Review of Psychology* 62: 363-390.

Herculano-Houzel, S., and R. Lent. (2009) "Isotropic Fractionator: A Simple, Rapid Method for the Quantification of Total Cell and Neuron Numbers in the Brain," *The Journal of Neuroscience* 25: 2518–521.

Hotz, Robert Lee. "Scientists Unveil Online Atlas of the Human Brain," *Wall Street Journal*, April 13, 2011.

Jabr, Ferris. "The Connectome Debate: Is Mapping the Mind of a Worm Worth It?" *Scientific American*, October 2, 2012.

Jain, V. et al. (2007) "Supervised Learning of Image Restoration with Convolutional Networks," *Proceedings: IEEE 11th International Conference on Computer Vision*.

Johnson, David. "Neurophilia," Guest Blog, *Scientopia*, July 23, 2011.

Johnson, S. C. et al. (2002) "Neural Correlates of Self-Reflection," *Brain* 125: 1808–814.

Kay, K. N. et al. (2008) "Identifying Natural Images from Human Brain Activity," *Nature* 452: 352–55.

Kelley, W. M. et al. (2002) "Finding the Self? An Event-Related fMRI Study," *Journal of Cognitive Neuroscience* 14: 785–94.

Koch, Christof. "The Connected Self," Nature.com, February 2, 2012.

———. "How Neuroscientists Observe Brains Watching Movies," *Scientific American*, December 29, 2012.

Kwong, K. K. et al. (1992) "Dynamic Magnetic Resonance Imaging of Human Brain Activity During Primary Sensory Stimulation," *Proceedings of the National Academy of Sciences* 89: 5675–679.

Lavater, J. C. *Essays on Physiognomy*. London: J. Murray, 1789.

LeDoux, Joseph. *The Synaptic Self: How Our Brains Become Who We Are*. New York: Penguin, 2003.

Littlefield, M. (2009) "Constructing the Organ of Deceit: The Rhetoric of fMRI and Brain Fingerprinting in Post-9/11 America," *Science, Technology & Human Values* 34: 365–92.

Madrigal, Alexis. "Scanning Dead Salmon in fMRI Machine Highlights Risk of Red Herrings," *Wired Science, Wired*, September 18, 2009.

Marcus, Gary. "Neuroscience Fiction," *The New Yorker*, December 2, 2012.

Miller, Greg. "Seeing Through the Mind's Eye," *Science Now, Science*, March 5, 2008.

Mitchell, J. P. et al. (2005) "The Link Between Social Cognition and Self-Referential Thought in the Medial Prefrontal Cortex," *Journal of Cognitive Neuroscience* 17: 1306–315.

Miyawaki, Y. et al. (2008) "Visual Image Reconstruction from Human Brain Activity Using a Combination of Multiscale Local Image Decoders," *Neuron* 60: 915–29.

O'Connor, C. et al. (2012) "Neuroscience in the Public Sphere," *Neuron* 74: 220–26.

Ogawa, S. et al. (1992) "Intrinsic Signal Changes Accompanying Sensory Stimulation: Functional Brain Mapping with Magnetic Resonance Imaging," *Proceedings of the National Academy of Sciences* 89: 5951–955.

Ouellette, Jennifer. "Tell Me No Lies," *Cocktail Party Physics, Scientific American*, June 16, 2012.

Penfield, W. (1924) "Oligodendroglia and Its Relation to Classic Neuroglia," *Brain* 47: 430–52.

Penfield, W., and E. Boldrey. (1937) "Somatic Motor and Sensory Representation in the Cerebral Cortex of Man as Studied by Electrical Stimulation," *Brain* 60: 349–443.

Penfield, W., and O. Foerster. (1930) "The Structural Basis of Traumatic Epilepsy and Results of Radical Operation," *Brain* 53: 8–119.

Penfield, Wilder, and Theodore Rasmussen. *The Cerebral Cortex of Man*. New York: Macmillan, 1950.

Plateka, S. M. et al. (2004) "Where Am I? The Neurological Correlates of Self and Other," *Cognitive Brain Research* 19: 114–22.

Poeppel, D. (2008) "The Cartographic Imperative: Confusing Localization and Explanation in Human Brain Mapping." In *Bildwelten des Wissens, "Ikonographie des Gehirns,"* edited by Bredekamp, Bruhn & Werner (Bildwelten volume 6.1, Spring 2008).

Quart, Alissa. "Neuroscience: Under Attack," *New York Times*, November 25, 2012.

Ramachandran, V. S. *The Tell-Tale Brain: A Neuroscientist's Quest for What Makes Us Human*. New York: W. W. Norton, 2011.

Randerson, James. "How Many Neurons Make a Human Brain? Billions Fewer Than We Thought," *Notes and Theories, The Guardian*, February 28, 2012.

Schiller, Francis. *Paul Broca: Explorer of the Brain*. Oxford, UK: Oxford University Press; 1992.

———. "Ig Nobel Prize in Neuroscience: The Dead Salmon Study," *The Scicurious Brain, Scientific American*, September 25, 2012.

Seung, Sebastian. *Connectome: How the Brain's Wiring Makes Us Who We Are*. New York: Houghton Mifflin Harcourt, 2012.

———. "Connectomics: Tracing the Wires of the Brain," The Dana Foundation, November 3, 2008.

Smith, Dana. "I Saw the Negative Sign: Problems with fMRI Research," *Brain Study,* July 10, 2012.

Smith, Kerri. "Brain Imaging: fMRI 2.0," Nature.com, April 4, 2012.

Stix, Gary. "Neuroscientists Stumble When They Make Conclusions from Examining Single Patients," *Observations, Scientific American*, March 30, 2012.

Turaga, S. C. et al. (2010) "Maximin Affinity Learning of Image Segmentation," *Advances in Neural Information Processing Systems* 22 (Proceedings of NIPS '09).

Thyreau, B. et al. (2012) "Very Large fMRI Study Using the IMAGEN Database: Sensitivity–Specificity and Population Effect Modeling in Relation to the Underlying Anatomy," *NeuroImage* 61: 295–303.

Uddin, L. Q. et al. (2006) "rTMS to the Right Inferior Parietal Lobule Disrupts Self–Other Discrimination," *Social Cognitive and Affective Neuroscience* 1: 65–71.

———. (2007) "The Self and Social Cognition: The Role of Cortical Midline Structures and Mirror Neurons," *Trends in Cognitive Sciences* 11: 153–157.

Van Horn, J. D. et al. (2012) "Mapping Connectivity Damage in the Case of Phineas Gage," *PLoS One* 7: e37454.

Vergano, Dan. "Blame It On the Brain: The Gray Matter of Politics," *USA Today*, May 22, 2006.

Voytek, Bradley. "How to Be a Neuroscientist," *Oscillatory Thoughts*, January 2011.

Weisberg, D. et al. (2008) "The Seductive Allure of Neuroscience Explanations," *Journal of Cognitive Neuroscience* 20: 470–77.

Westly, Erica. "Brain Imaging Reveals What You're Watching," *Technology Review*, September 22, 2011.

Witchalls, Clint. "Murder in Mind," *The Guardian*, March 24, 2004.

Yong, Ed. "Will We Ever Have a Foolproof Lie Detector?" *Not Exactly Rocket Science*, *Discover*, April 9, 2012.

Zimmer, Carl. "From Cooling System to Thinking Machine," *Being Human*, October 10, 2012.

———. "Once We Model the Connectome, We'll Glimpse the Anatomy of the Mind," *Discover*, April 2012.

Online Resources

The Great Brain Mapping Debate: http://www.youtube.com/watch?v=fRHzk RqGf-g

Sebastian Seung TED talk: http://www.ted.com/talks/sebastian_seung.html

Wilder Penfield video: http://www.youtube.com/watch?v=kNdM9JhTPJw

3.   Moveable Types

Addams Family Values. (1993) Paramount. Dir. Barry Sonnenfeld. Screenplay: Paul Rudnick.

Allport, G. W. *Personality: A Psychological Interpretation*. New York: Holt, 1937.

Allport, G. W., and H. S. Odbert. (1936) "Trait Names: A Psychological Study," *Psychological Monographs* 47: 211.

Asendorpf, J. B. (2002) "The Puzzle of Personality Types," *European Journal of Personality* 16: S1–S6.

Back, M. D. et al. (2009) "Predicting Actual Behavior from the Explicit and Implicit Self-Concept of Personality," *Journal of Personality and Society Psychology* 97: 533–48.

Bemmels, H. R. et al. (2008) "The Heritability of Life Events: An Adolescent Twin and Adoption Study," *Twin Research in Human Genetics* 11: 257–65.

Benjamin, J. et al. (1996) "Population and Familial Association Between the D4 Dopamine Receptor Gene and Measures of Novelty Seeking," *Nature Genetics* 12: 81–84.

Bouchard, T. J. Jr. (1993) "Genes, Environment, and Personality," *Science* 264: 1700–701.

———. (2004) "Genetic Influence on Human Psychological Traits: A Survey," *Current Directions in Psychological Science* 13: 148–51.

Bouchard, Thomas J., Jr., and M. McGue. (2003) "Genetic and Environmental Influences on Human Psychological Differences," *Journal of Neurobiology* 54: 4–45.

Bouchard, T. J. Jr. et al. (1990) "Sources of Human Psychological Differences: The Minnesota Study of Twins Reared Apart," *Science* 250: 223–28.

Burton, Robert. *The Essential Anatomy of Melancholy*. Mineola, NY: Dover Books, 2002.

Canli, T. (2004) "Functional Brain Mapping of Extraversion and Neuroticism: Learning from Individual Differences in Emotion Processing," *Journal of Personality* 72: 1105–131.

Caspi, A. et al. (2003) "Influence of Life Stress on Depression: Moderation by a Polymorphism of the 5-HTT Gene," *Science* 301: 386–89.

Claridge, J., and C. Davis. (2001) "What's the Use of Neuroticism," *Personality and Individual Differences* 31: 383–400.

Cloninger, C. R. et al. (1996) "Mapping Genes for Human Personality," *Nature Genetics* 12: 3–4.

Cravchick, A., and D. Goldman. (2000) "Neurochemical Individuality: Genetic Diversity among Human Dopamine and Serotonin Receptors and Transporters," *Archives of General Psychiatry* 57: 1105–114.

Depue, R. A., and P. F. Collins. (1999) "Neurobiology of the Structure of Personality: Dopamine, Facilitation of Incentive Motivation, and Extraversion," *Behavioral and Brain Sciences* 22: 491–569.

Depue, R. A. et al. (1994) "Dopamine and the Structure of Personality: Relation of Agonist-Induced Dopamine Activity to Positive Emotionality," *Journal of Personality and Social Psychology* 67: 485–98.

Dickson, D. H., and I. W. Kelly. (1985) "'The Barnum Effect' in Personality Assessment: A Review of the Literature," *Psychological Reports* 57: 367–82.

Dobbs, David. "The Science of Success," *The Atlantic*, December 2009.

Ebstein, R. (2006) "The Molecular Genetic Architecture of Human Personality: Beyond Self-Report Questionnaires," *Molecular Psychiatry* 11: 427–45.

Ebstein, R. et al. (1995) "Dopamine D4 Receptor (D4DR) Exon III Polymorphism Associated with the Human Personality Trait of Novelty Seeking," *Nature Genetics* 12: 78–80.

Fidler, A. E. et al. (2007) "DRD4 Gene Polymorphisms Are Associated with Personality Variation in a Passerine Bird," *Proceedings of the Royal Society B* 274: 1685–691.

Gage, Fred H., and Alysson R. Muotri. "What Makes Each Brain Unique?" *Scientific American*, March 2012.

Galton, Francis. (1884) "The Measurement of Character," *Fortnightly Review* 36: 179–85.

———. (1885) "The Measurement of Fidget," *Nature* 32: 174–75.

Hamer, Dean, and Peter Copeland. *Living with Our Genes: The Groundbreaking Book About the Science of Personality, Behavior, and Genetic Destiny*. New York: Anchor Books, 1999.

Hess, Amanda. "Dry T-Shirt Contest: I Sniffed Armpit Stains at a Pseudoscientific Singles Pheromone Party," *GOOD*, April 9, 2012.

Higgins, E. T. (1987). "Self-Discrepancy: A Theory Relating Self and Affect," *Psychological Review* 94: 319–40.

Holmes, Hannah. *Quirk: Brain Science Makes Sense of Your Peculiar Personality*. New York: Random House, 2011.

Horn, N. R. et al. (2003) "Response Inhibition and Impulsivity: an fMRI Study," *Neuropsychologia* 41: 1959–966.

Howes, R. J., and T. G. Carskadon. (1979) "Test-Retest Reliabilities of the Myers-Briggs Type Indicator as a Function of Mood Changes," *Research in Psychological Type* 2: 67–72.

Jang, K. L. et al. (1996) "Heritability of the Big Five Personality Dimensions and Their Facets: A Twin Study," *Journal of Personality* 64: 577–91.

———. (1998) "Heritability of Facet-Level Traits in Cross-Cultural Twin Sample: Support for a Hierarchical Model of Personality," *Journal of Personality and Social Psychology* 74: 1556–575.

John, O. P. "The Big Five Factor Taxonomy: Dimensions of Personality in the Natural Language and Questionnaires." In *Handbook of Personality: Theory and Research,* Lawrence A. Pervin, ed. New York: Guilford Press, 1990, 66–100.

Johnson, A. et. al. "Behavioral Genetic Studies of Personality: An Introduction and Review of the Results of 50+ Years of Research." In *Handbook of Personality: Theory and Assessment,* Gregory Boyle, Gerald Matthews, and Donald Saklofske, eds. London: Sage Publishers, 2008, 145–73.

Johnson, W. (2007) "Genetic and Environmental Influences on Behavior: Capturing All the Interplay," *Psychology Review* 114: 423–40.

Jung, Carl G. *Psychological Types.* New York: Harcourt Brace, 1921.

Kagan, J. et al. (1988) "Biological Bases of Childhood Shyness," *Science* 240: 167–71.

Kagan, Jerome. *Galen's Prophecy: Temperament in Human Nature.* New York: Basic Books, 1994.

Korsten, P. et al. (2010) "Association Between DRD4 Gene Polymorphism and Personality Variation in Great Tits: a Test Across Four Wild Populations," *Molecular Ecology* 19: 832–43.

LaHaye, Tim. *Spirit Controlled Temperament.* Revised edition. Carol Stream, IL: Tyndale House Publishers, 1994.

Lesch, K-P. et al. (1996) "Association of Anxiety-Related Traits with a Polymorphism in the Serotonin Transporter Gene Regulatory Region," *Science* 274: 1527–531.

Lynch, Michael, and Bruce Walsh. *Genetics and Analysis of Quantitative Traits.* Sunderland, MA: Sinauer, 1998.

McCrae, Robert R. (1982) "Consensual Validation of Personality Traits: Evidence from Self-Reports and Ratings," *Journal of Personality and Social Psychology* 43: 293–303.

McCrae, R. R., and O. P. John. (1992) "An Introduction to the Five-Factor Model and Its Applications," *Journal of Personality* 60: 175–215.

McGue, M. et al. (1993) "Personality Stability and Change in Early Adulthood: A Behavioral Genetic Analysis," *Developmental Psychology* 29: 96–109.

Mischel, Walter. *Personality and Assessment.* New York: Wiley, 1968.

Mischel, W., and Y. Shoda. (1995) "A Cognitive-Affective System Theory of Personality: Reconceptualizing Situations, Dispositions, Dynamics and Invariance in Personality Structure," *Psychology Review* 102: 246–68.

Munafo, M. R. et al. (2003) "Genetic Polymorphisms and Personality in Healthy Adults: A Systematic Review and Meta-Analysis," *Molecular Psychiatry* 8: 471–84.

Nettle, Daniel. *Personality: What Makes You the Way You Are.* Oxford, UK: Oxford University Press, 2007.

Nettle, D., and L. Penke. (2010) "Personality: Bridging the Literatures from Human Psychology and Behavioral Ecology," *Philosophical Transactions of the Royal Society B* 365: 4043–50.

Nuwer, Rachel. "You Are What You Bleed: In Japan and Other East Asian Countries, Some Believe Blood Type Dictates Personality," Guest Blog, *Scientific American*, February 15, 2011.

Omura, K. et al. (2005) "Amygdala Gray Matter Concentration Is Associated with Extraversion and Neuroticism," *Neuroreport* 16: 1905–908.

Paul, Annie Murphy. *The Cult of Personality Testing.* New York: Free Press, 2004.

Pedersen, N. L. et al. (1988) "Neuroticism, Extraversion, and Related Traits in Adult Twins Reared Apart and Together," *Journal of Personality and Social Psychology* 55: 950–57.

Peterson, J. B., and S. Carson. (2000) "Latent Inhibition and Openness in a High-Achieving Student Population," *Personality and Individual Difference* 28: 323–32.

Poulin, M. J. et al. (2012) "The Neurogenetics of Nice: Receptor Genes for Oxytocin and Vasopressin Interact with Threat to Predict Prosocial Behavior," *Psychological Science* 23: 446–52.

Roberts, B. W., and D. K. Mroczek. (2008) "Personality Trait Stability and Change," *Current Directions in Psychological Science* 17: 31–5.

Roberts, B. W. et al. (2007) "The Power of Personality: the Comparative Validity of Personality Traits, Socio-Economic Status and Cognitive Ability for Predicting Important Life Outcomes," *Perspectives on Psychological Science* 2: 313–45.

Schmidt, L. A., and N. A. Fox. (1995) "Individual Differences in Young Adults' Shyness and Sociability: Personality and Health Correlates," *Personality and Individual Differences* 19: 455–62.

Schneider, T. R. (2004) "The Role of Neuroticism on Psychological and Phys-
    iological Stress Responses," *Journal of Experimental Social Psychology*
    40: 795–804.

Soldz, S., and G. E. Vaillant. (1999) "The Big Five Personality Traits and the
    Life Course: A 45-Year Longitudinal Study," *Journal of Research in Per-
    sonality* 33: 208–32.

Sorokowska, A. et al. (2011) "Does Personality Smell? Accuracy of Personal-
    ity Assessments Based on Body Odor," *European Journal of Personality*
    26: 496–503.

Swann, W. B. Jr., (1997) "The Trouble with Change: Self-Verification and
    Allegiance to the Self," *Psychological Science* 8: 177–80.

Swann, W. B. Jr., et al. (1989) "Agreeable Fancy or Disagreeable Truth? Rec-
    onciling Self-Enhancement and Self-Verification," *Journal of Personality
    and Social Psychology* 57: 782–91.

Tannenbaum, Melanie. "Anything But Country: What Factor Analysis Re-
    veals About Our Taste for Tunes," Guest Blog, *Scientific American*, Au-
    gust 8, 2011.

Tellegen, A. et al. (1988) "Personality Similarity in Twins Reared Apart and
    Together," *Journal of Personality and Social Psychology* 54: 1031–39.

Trilling, Lionel. "The Freud/Jung Letters," *The New York Times*, April 21,
    1974.

Vazire, S., and M. R. Mehl. (2008) "Knowing Me, Knowing You: The Rela-
    tive Accuracy and Unique Predictive Validity of Self-Ratings and Other-
    Ratings of Daily Behavior," *Journal of Personality and Social Psychology*
    95: 1202–216.

Vitelli, Romeo. "What's Your Blood Type?" *Providentia*, August 23, 2011.

Vul, E. et al. (2009) "Puzzlingly High Correlations in fMRI Studies of Emo-
    tion, Personality and Social Cognition," *Perspectives on Psychological
    Science* 4: 274–90.

Yamagata, S. et al. (2006) "Is the Genetic Structure of Human Personality
    Universal? A Cross-Cultural Twin Study from North America, Europe
    and Asia," *Journal of Personality and Social Psychology* 90: 987–98.

Online Resources

BBC on Anatomy of Melancholy: http://www.bbc.co.uk/programmes/
    b010y30m

The Personality Project: http://personality-project.org/perproj/readings
    -theory.html

II. MYSELF

4. Three and I'm Under the Table

Abbot, Karen. "The Man Who Wouldn't Die," *Smithsonian,* February 7, 2012.

Akst, Jeff. "Binge-Drinking Mice," *The Scientist,* December 12, 2011.

Allison, M. et al. (2011) "Alcohol Intake in Prairie Voles Is Influenced by the Drinking Level of a Peer," *Alcoholism: Clinical and Experimental Research* 35: 1884–890.

Bachtell, R. K. et al. (2003) "The Edinger-Westphal-Lateral Septum Urocortin Pathway and Its Relationship to Alcohol Consumption," *Journal of Neuroscience* 23: 2477–487.

Bainton, R. J. et al. (2000) "Dopamine Modulates Acute Responses to Cocaine, Nicotine, and Ethanol in Drosophila," *Current Biology* 10: 187–94.

Belknap, J. K., and A. L. Atkins (2001) "The Replicability of QTLs for Murine Alcohol Preference Drinking Behavior Across Eight Independent Studies," *Mammalian Genome* 12: 893–99.

Brink, Susan. "Your Brain on Alcohol," *US News,* April 29, 2001.

Brookshire, Bethany. "Impulsivity, Addiction and Your Synapses," *The Scicurious Brain, Scientific American,* November 23, 2011.

Buckholtz, J. W. et al. (2010) "Dopaminergic Network Differences in Human Impulsivity," *Science* 329: 532.

Cadoret, R. J. et al. (1984) "Alcoholism and Antisocial Personality: Interrelationships, Genetic and Environmental Factors," *Archives of General Psychiatry* 42: 161–67.

———. (1979) "Development of Alcoholism in Adoptees Raised Apart from Alcoholic Biologic Relatives," *Archives of General Psychiatry* 37: 561–63.

Cardenas, V.A. et al. (2011) "Brain Morphology at Entry into Treatment for Alcohol Dependence Is Related to Relapse Propensity," *Biological Psychiatry* 70: 61–7.

Casey, B. J. et al. (2011) "Behavioral and Neural Correlates of Delay of Gratification 40 Years Later," *Proceedings of the National Academy of Sciences* 108: 14998–5003.

Chester, J. A. et al. (2002). "High and Low Alcohol-Preferring Mice Show Differences in Conditioned Taste Aversion to Alcohol," *Alcoholism: Clinical and Experimental Research* 27: 12–18.

Cloninger, C. R. et al. (1981) "Inheritance of Alcohol Abuse: Cross-Fostering Analysis of Adopted Men," *Archives of General Psychiatry* 36: 861–68.

Corl, A. et al. (2005) "Insulin Signaling in the Nervous System Regulates Ethanol Intoxication in Drosophila," *Nature Neuroscience* 8: 8–9.

Cotton, N. S. (1979) "The Familial Incidence of Alcoholism: A Review," *Journal of Studies on Alcoholism* 40: 89–116.

De Quincey, Thomas. *Confessions of an English Opium-Eater and Other Writings.* New York: Penguin Classics, 2003.

Dick, Danielle M., and Tatiana Foroud. "Genetic Strategies to Detect Genes Involved in Alcoholism and Alcohol-Related Traits," National Institute on Alcohol Abuse and Alcoholism, June 2003.

Dirda, Michael. "The Fascinating Life of an English Writer, Essayist, and 'Opium Eater,'" *The Washington Post*, December 30, 2010.

Durazzo, T. C. et al. (2008) "Combined Neuroimaging, Neurocognitive and Psychiatric Factors to Predict Alcohol Consumption Following Treatment for Alcohol Dependence," *Alcohol* 43: 683–91.

Eigste, I-M. et al. (2006) "Predicting Cognitive Control from Preschool to Late Adolescence and Young Adulthood," *Psychological Science* 17: 478–84.

Ferguson, Craig. *American on Purpose: The Improbable Adventures of an Unlikely Patriot.* New York: Harper, 2009.

Friedman, Richard A. "Who Falls to Addiction and Who Is Unscathed?" *The New York Times*, August 1, 2011.

Gage, Suzi. "Have Scientists Discovered a Gene for Binge-Drinking? No, But the Research Is Still Important," *Scilogs,* December 5, 2012.

Gazdzinski, S. et al. (2008) "Are Treated Alcoholics Representative of the Entire Population with Alcohol Use Disorders? A Magnetic Resonance Study of Brain Injury," *Alcohol* 42: 67–76.

———. (2010) "Cerebral White Matter Recovery in Abstinent Alcoholics—A Multimodality Magnetic Resonance Study," *Brain* 133: 1043–53.

Grahame, N. J. (2000) "Hunt for the Genes Involved in Alcoholism: Strain Differences and Selective Breeding," *Alcohol Health and Research World* 23: 159–63.

Grahame, N. J., and C. L. Cunningham. (2002) "Intravenous Self-Administration of Ethanol in the Mouse," *Current Protocols in Neuroscience Supplement* 19, Unit 9.11.1–9.11.19.

Grahame, N. J. et al. (2000) "Ethanol Locomotor Sensitization, but not Tolerance Correlates with Selection for Alcohol Preference in High- and Low-Alcohol Preferring Mice," *Psychopharmacology* 151: 252–60.

Haughey, H. M. et al. (2008) "Human Gamma-Aminobutyric Acid A Receptor Alpha2 Gene Moderates the Acute Effects of Alcohol and Brain mRNA Expression," *Genes Brain Behavior* 7: 447–54.

Klein, Debra. "Binge Drinking May Hamper Information Relay System in Teen Brain," *University of San Diego News*, April 22, 2009.

Koerner, Brendan. "Secret of AA: After 75 Years, We Don't Know How It Works," *Wired*, June 23, 2010.

Koob, G. F. (2003) "Alcoholism: Allostasis and Beyond," *Alcoholism: Clinical and Experimental Research* 27: 232–43.

Koob, G. F. et al. (1998) "Neurocircuitry Targets in Ethanol Reward and Dependence," *Alcoholism: Clinical and Experimental Research* 22: 3–9.

———. (1998) "Neuroscience of Addiction," *Neuron* 21: 467–76.

Le Stret, Y. et al. (2008) "The 3' Part of the Dopamine Transporter Gene DAT1/SLC6A3 Is Associated with Withdrawal Seizures in Patients with Alcohol Dependence," *Alcoholism: Clinical and Experimental Research* 32: 27–35.

Li, D. et al. (2011) "Strong Protective Effect of the Aldehyde Dehydogenase Gene (ALDH2) 504lys (*2) Allele Against Alcoholism and Alcohol-Induced Medical Diseases in Asians," *Human Genetics* 131: 725–37.

McGue, M. (1999) "The Behavioral Genetics of Alcoholism," *Current Directions in Psychological Science* 8: 109–55.

Mischel, W. et al. (1972) "Cognitive and Attentional Mechanisms in Delay of Gratification," *Journal of Personality and Social Psychology* 21: 204–18.

Moore, M. S. et al. (1998) "Ethanol Intoxication in Drosophila: Genetic and Pharmacological Evidence for Regulation by the cAMP Pathway," *Cell* 93: 997–1007.

Morrison, Robert. *The English Opium Eater: A Biography of Thomas de Quincey.* New York: Phoenix, 2010.

Munafo, M. R. et al. (2007) "Association of the DRD2 gene Taq1A Polymorphism and Alcoholism: A Meta-Analysis of Case-Control Studies and Evidence of Publication Bias," *Molecular Psychiatry* 12: 454–61.

Murray, R. M., and C. C. Gurlin. "Twin and Alcoholism Studies." In *Recent Developments in Alcoholism*, Vol. 1, Marc Galanter, ed. New York: Gardner Press, 1983, 25–48.

Partanen, J. et al. "Inheritance of Drinking Behavior: A Study on Intelligence, Personality and Use of Alcohol in Adult Twins." In *Emerging Concepts of Alcohol Dependence*. New York: Springer Publishing, Inc., 1977.

Pearson, Edmund. "Malloy the Mighty," *The New Yorker*, September 23, 1933.

Prescott, C. A. et al. (2005) "The Washington University Twin Study of Alcoholism," *American Journal of Medical Genetics Part B (Neurophysics Genetics)* 134B: 48–55.

Ray, L. A. et al. (2009) "The Dopamine D Receptor (DRD4) Gene Exon III Polymorphism, Problematic Alcohol Use and Novelty Seeking: Direct and Mediated Genetic Effects," *Addiction Biology* 14: 238–44.

———. (2009) "Effects of Naltrexone on Cortisol Levels in Heavy Drinkers," *Pharmacolology Biochemical Behavior* 91: 489–94.

———. (2009) "A Preliminary Pharmacogenetic Investigation of Adverse Events from Topiramate in Heavy Drinkers," *Experimental Clinical Psychopharmacology* 17: 122–29.

Read, Simon. *On the House: The Bizarre Killing of Michael Malloy.* New York: Berkeley Books, 2005.

Rodan, A. R. et al. (2002) "Functional Dissection of Neuroanatomical Loci Regulating Ethanol Sensitivity in Drosophila," *Journal of Neuroscience* 22: 9490–501.

Rothenfluh, A. et al. (2006) "Distinct Behavioral Responses to Ethanol Are Regulated by Alternate RhoGAP18B Isoforms," *Cell* 127: 199–211.

Scholz, H. et al. (2000) "Functional Ethanol Tolerance in Drosophila," *Neuron* 28: 261–71.

———. (2005) "The Hangover Gene Defines a Stress Pathway Required for Ethanol Tolerance Development," *Nature* 436: 845–47.

Schuckit, M. A. (2009) "An Overview of Genetic Influences in Alcoholism," *Journal of Substance Abuse Treatment* 36: S5–S14.

Shoda, Y. et al. (1990) "Predicting Adolescent Cognitive and Self-Regulatory Competencies from Preschool Delay of Gratification: Identifying Diagnostic Conditions," *Developmental Psychology* 26: 978–86.

Stacey, D. et al. (2012) "RASGRF-2 Regulates Alcohol-Induced Reinforcement by Influencing Mesolimbic Dopamine Neurone Activity and Dopamine Release," *Proceedings of the National Academy of Sciences* 109: 21128–133.

Tapert, Susan. "What Does Alcohol Do To Your Brain?" *Addiction Science, Psychology Today*, October 2, 2008.

Tapert, Susan, and Lindsay Squeglia. "Adolescent Binge Drinking Linked to Spatial Working Memory Brain Activation: Differential Gender Effects," *Alcoholism: Clinical and Experimental Research* 35: 1831–841.

Treutlein, J. et al. (2009) "Genome-Wide Association Study of Alcohol Dependence," *Archives of General Psychiatry* 66: 773–84.

Umemori, H. et al. (2011) "Multiple Forms of Activity-Dependent Competition Refine Hippocampal Circuits In Vivo," *Neuron* 70: 1128–142.

Vaillant, George. *The Natural History of Alcoholism Revisited*. Cambridge, MA: Harvard University Press, 1995.

———. (2003) "A 60-Year Follow-Up of Alcoholic Men," *Addiction* 98: 1043–51.

Wolf, F. W. et al. (2002) "High Resolution Analysis of Ethanol-Induced Locomotor Stimulation in Drosophila," *Journal of Neuroscience* 22: 11035–1044.

Yong, Ed. "Abnormal Structures Hint at Poor Self Control and Vulnerability to Drug Addiction," *Not Exactly Rocket Science, Discover*, February 2, 2012.

Zimbardo, Philip, and John Boyd. *The Time Paradox: The New Psychology of Time That Will Change Your Life*. New York: Free Press, 2008.

Online Resources

23andMe Web site: http://www.23andme.com

Ulrike Heberlein's drunken fruit flies: http://vimeo.com/8103538

Heberlein profile: http://www.youtube.com/watch?v=1HnRw5IyLVc

Walter Mischel's marshmallow test: http://www.youtube.com/watch?v=rMk n4J_l9uU

Philip Zimbardo lecture: http://www.youtube.com/watch?v=y7t-HxuI17Y

Joachim de Posada TED talk: http://www.youtube.com/watch?v=ykLUZO_ -QZk

Craig Ferguson on his alcoholism: http://www.youtube.com/watch?v=7ZV WIELHQQY

National Institute on Alcohol Abuse and Alcoholism http://www.niaaa.nih .gov

5.   My So-Called Second Life

Abcarian, Robin. "Celeb Note to Self: You Are Fabulous: A Scientific Study Shows That Stars Really are Narcissists First," *Los Angeles Times*, September 12, 2006.

Asimina V., and A. N. Joinson. (2009) "Me, Myself and I: The Role of Interactional Context on Self-Presentation Through Avatars," *Computers in Human Beha*vior 25: 510–20.

Back, M. D. et al. (2010) "Facebook Profiles Reflect Actual Personality Not Self-Idealization." *Psychological Science* 21: 372–74.

Bainbridge, W. S. (2007) "The Scientific Research Potential of Virtual Worlds," *Science* 317: 472–76.

Bak, P. et al. (1987) "Self-Organized Criticality: An Explanation of 1/f Noise," *Physical Review Letters* 59: 381–84.

Bargh, J. A. et al. (2002) "Can You See the Real Me? Activation and Expression of the 'True Self' on the Internet," *Journal of Social Issues* 58: 33–48.

Bélisle, J. F., and H. O. Bodur. (2010) "Avatars as Information: Perception of Consumers Based on Their Avatars in Virtual Worlds," *Psychology & Marketing* 27: 741–65.

Belk, R. W. (1988) "Possessions and the Extended Self," *Journal of Consumer Research* 15: 139–68.

Bessiere, K. et al. (2007) "The Ideal Elf: Identity Exploration in World of Warcraft," *CyberPsychology and Behavior* 10: 130–35.

Blascovich, Jim, and Jeremy Bailenson. *Infinite Reality: Avatars, Eternal Life, New Worlds, and the Dawn of the Virtual Revolution.* New York: William Morrow, 2011.

Botvinick, Matthew. (2004) "Probing the Neural Basis of Body Ownership," *Science* 305: 782–783.

Botvinick, M., and J. Cohen. (1998) "Rubber Hands 'Feel' Touch that Eyes See," *Nature* 391: 756.

Botvinick, M. et al. (2005) "Viewing Facial Expressions of Pain Engages Cortical Areas Engaged in the Direct Experience of Pain," *NeuroImage* 25: 312–19.

Brookshire, Bethany. "Mirror, Mirror on My Facebook Wall," *Neurotic Physiology, Scientopia,* February 28, 2011.

Burroughs, J. W. et al. (1991) "Predicting Personality from Personal Possessions: A Self-Presentational Analysis," *Journal of Social Behavior and Personality* 6: 147–53.

Csikszentmihalyi, Mihaly. "Why We Need Things," *History from Things*. Steven Lubar and W. David Kingery, eds. Washington D.C. Smithsonian Institution Press, 1993.

Demos, K. E. et al. (2008) "Human Amygdala Sensitivity to the Pupil Size of Others," *Cerebral Cortex* 18: 2729–734.

Dotov, D. G. et al. (2010) "A Demonstration of the Transition from Ready-to-Hand to Unready-to-Hand." *PLoS One* 5: e9433.

Dutta, P., and Horn, P.M. (1981) "Low-Frequency Fluctuations in Solids: 1/f Noise," *Reviews of Modern Physics* 53: 497–516.

Duval, T. S., and R. A. Wicklund. *A Theory of Objective Self-Awareness*. New York: Academic, 1972.

———. (1973) "Effects of Objective Self-Awareness on Attributions of Causality," *Journal of Experimental Social Psychology* 9: 17–31.

Ehrsson, H. H. et al. (2004) "That's My Hand! Activity in Premotor Cortex Reflects Feeling of Ownership of a Limb," *Science* 305: 875–77.

Fron, J. et al. (2007) "Playing Dress-Up: Costume, Roleplay and Imagination," *Philosophy of Computer Games Online Proceedings*, January 2007.

Goffman, Erving. *The Presentation of Self in Everyday Life*. Garden City, NY: Doubleday, 1959.

Gonzales, A., and J. Hancock. (2011) "Mirror, Mirror on My Facebook Wall: Effects of Exposure to Facebook on Self-Esteem," *Cyberpsychology, Behavior, and Social Networking* 14: 79–83.

Gosling, Sam. *Snoop: What Your Stuff Says About You*. New York: Basic Books, 2008.

Gosling, S. D. et al. (2011) "Manifestations of Personality in Online Social Networks: Self-Reported Facebook-Related Behaviors and Observable Profile Information," *Cyberpsychology, Behavior, and Social Networking* 14: 483–88.

———. (2007) "Personality Impressions Based on Facebook Profiles," *Proceedings of the International Conference on Weblogs and Social Media*, Boulder, CO, March 26 28, 2007.

————. (2002) "A Room with a Cue: Personality Judgments Based on Offices and Bedrooms," *Journal of Personality and Society Psychology* 82: 379–98.

Hess, E. H. (1965) "Attitude and Pupil Size," *Scientific American* 212: 46–54.

Hoffman, H.G. (2004) "Virtual Reality Therapy," *Scientific American*, July 26, 2004.

Hussain, Z., and M. D. Griffiths. (2008) "Gender Swapping and Socializing in Cyberspace: An Exploratory Study," *CyberPsychology and Behavior* 11: 47–53.

Jablonski, Chris. "A Better Wearable Brain-Computer Interface," ZDNet .com, August 16, 2011.

Janssen, J. H. et al. "Intimate Heartbeats: Opportunities for Affective Communications Technology," *IEEE Transactions in Affective Computing* 1: 72–80.

Kepplinger, H. M., and W. Donsbach. (1990) "The Impact of Camera Perspectives on the Perception of a Speaker," *Communication* 12: 38–43.

Kilteni, K. et al. (2012) "Extending Body Space in Immersive Virtual Reality: A Very Long Arm Illusion," *PLoS One* 7: e40867.

Kim, D-H. et al. (2011) "Epidermal Electronics," *Science* 333: 838–43.

Konijn, A., and M. N. Bijvank. "Door to Another Me: Identity Construction Through Digital Game Play." In *Serious Games: Mechanisms and Effects*. Ute Ritterfeld, Michael Cody, and Peter Vorderer, eds. New York: Routledge, 2009, 179–203.

Krossa, E. et al. (2011) "Social Rejection Shares Somatosensory Representations with Physical Pain," *Proceedings of the National Academy of Sciences*. 108: 6270–6275.

Lacan, J. "Some Reflections on the Ego." In *Ecris: The First Complete Edition in English*. Translated by Bruce Fink. New York: W. W. Norton, 2006.

McKeon, Matt, and Susan Wyche. *Life Across Boundaries: Identity and Gender in Second Life*. Atlanta, GA: Georgia Institute of Technology, 2006.

Michaels, Walter Benn. *The Shape of the Signifier: 1967 to the End of History*. Princeton, NJ: Princeton University Press, 2004.

Morie, J. F. (2008) "The Performance of the Self and Its Effect on Presence in Virtual Worlds," *Proceedings of the 11th International Workshop on Presence*, Padua, Italy, 265–69.

————. (2007) "Performing in (Virtual) Spaces: Embodiment and Being in Virtual Environments," *International Journal of Performing Arts and Digital Media* 3: 123–38.

Moseley, G. L. et al. (2008). "Psychologically Induced Cooling of a Specific Body Part Caused by the Illusory Ownership of an Artificial Counterpart," *Proceedings of the National Academy of Sciences* 105: 13169–3173.

Neustaedter, C., and E. Fedorovskaya. (2009) "Presenting Identity in a Virtual World Through Avatar Appearances," *Proceedings of Graphics Interface*, ACM Press.

Nowak, K. L., and C. Rauh. (2008) "Choose Your 'Buddy Icon' Carefully: The Influence of Avatar Androgyny, Anthropomorphism and Credibility in Online Interactions," *Computers and Human Behavior* 24: 1473–493.

O'Brien, L., and J. Murnane. (2009) "An Investigation into How Avatar Appearance Can Affect Interactions in a Virtual World," *International Journal of Social and Humanistic Computing* 1: 192–202.

Ouellette, Jennifer. "Writer Jennifer Ouellette on Her Twitter Handle, JenLucPiquant," *The Atlantic (online)*, August 11, 2011.

Partala, T. (2011) "Psychological Needs and Virtual Worlds: Case Second Life," *International Journal of Human-Computer Studies* 69: 787–800.

Petkova, V. I., and H. H. Ehrsson. (2008) "If I Were You: Perceptual Illusion of Body Swapping," *PLoS One* 3: e3832.

Prelinger, E. (1959) "Extension and Structure of the Self," *Journal of Psychology* 47: 13–23.

Rowling, J. K. *Harry Potter and the Sorcerer's Stone.* New York: Scholastic, 1998.

Ryan, R. M., and E. L. Deci. (2000) "Self-Determination Theory and the Facilitation of Intrinsic Motivation, Social Development, and Well-Being," *American Psychologist* 55: 68–78.

Schroeder, Ralph (ed). *The Social Life of Avatars.* London: Springer-Verlag, 2002.

Silvia, P. J., and S. T. Duval. (2001) "Objective Self-Awareness Theory: Recent Progress and Enduring Problems," *Personality and Social Psychology Review* 5:230–41.

Stephenson, Neal. *Snow Crash.* New York: Bantam, 2003.

Thakkar, N. et al. (2011) "Disturbances in Body Ownership in Schizophrenia: Evidence from the Rubber Hand Illusion and Case Study of a Spontaneous Out-of-Body Experience," *PLoS One* 6: e27089.

Trepte, S., and L. Reinecke. (2010) "Avatar Creation and Video Game Enjoyment: Effects of Life-Satisfaction, Game Competitiveness, and Identification with the Avatar," *Journal of Media Psychology* 22: 171–84.

Turkle, S. (1994) "Construction and Reconstruction of Self in Virtual Reality: Playing in the MUDs," *Mind, Culture, and Activity: An International Journal* 1: 158–67.

Vasalou, A., and A. Joinson. (2009) "Me, Myself and I: The Role of Interactional Context on Self-Presentation Through Avatars," *Computers in Human Behavior* 25: 510–20.

Vazire, S., and S. D. Gosling. (2004) "e-Perceptions: Personality Impressions Based on Personal Web sites," *Journal of Personality and Social Psychology* 87: 123–32.

Vezina, Kenrick. "Stick-on Electronic Tattoos," *Technology Review*, August 11, 2011.

Walther, J. B. (1992) "Interpersonal Effects in Computer-Mediated Communication: A Relational Perspective," *Communication Research* 19: 52–90.

———. (2007) "Selective Self-Presentation in Computer-Mediated Communication: Hyperpersonal Dimensions of Technology, Language and Cognition," *Computers in Human Behavior* 23: 2538–557.

Webb, Eugene, Donald Campbell, Richard Schwaste, and Lee Seacrest. *Unobtrusive Measures: Non-reactive Research in the Social Sciences.* Chicago, IL: Rand McNally, 1966.

Weibel, D. et al. (2010) "How Socially Relevant Visual Characteristics of Avatars Influence Impression Formation," *Journal of Media Psychology* 22: 37–43.

Yee, N., and J. N. Bailenson. (2007) "The Proteus Effect: The Effect of Transformed Self-Representation on Behavior," *Human Communication Research* 33: 271–90.

Yee, N. et al. (2011) "The Expression of Personality in Virtual Worlds," *Social Psychological and Personality Science* 2: 5–12.

———. (2007) "The Unbearable Likeness of Being Digital: The Persistence of Nonverbal Social Norms in Online Virtual Environments," *CyberPsychology and Behavior* 10: 115–21.

Yong, Ed. "Out-of-Body Experience: Master of Illusion," Nature.com, December 7, 2011.

Online Resources

Tony Chemero's lecture: http://vimeo.com/33552945

Alexandra Samuel's TEDx talk: "Virtual Life Is Real Life": http://youtu.be/ui2ZwO-efo0

Lanier, Jared. "Homuncular Flexibility," Edge World Question Center, 2006: http://www.edge.org/q206/q06_7.html#lanier

6.   Born This Way

Adams, William Lee. "Could This Women's World Champ Be a Man?" *Time*, August 21, 2009.

Alexander, G. M. (2003) "An Evolutionary Perspective of Sex-Typed Toy Preferences: Pink, Blue, and the Brain," *Archives of Sexual Behavior* 32: 7–14.

Arana, Gabriel. "My So-Called Ex-Gay Life," *The American Prospect*, April 11, 2012.

Arnold, A. P., and S. M. Breedlove. (1985) "Organizational and Activational Effects of Sex Steroids on Brain and Behavior: a Reanalysis," *Hormones and Behavior* 19: 469–98.

Auyeung, B. et al. (2009) "Fetal Testosterone Predicts Sexually Differentiated Childhood Behavior in Girls and in Boys," *Psychological Science* 20: 144–48.

Bagemihl, Bruce. *Biological Exuberance: Animal Homosexuality and Natural Diversity*. New York: St. Martin's Press, 1999.

Bailey, J. M., and R. C. Pillard. (1991) "A Genetic Study of Male Sexual Orientation," *Archives of General Psychiatry* 48: 1089–96.

Bailey, J. M., and K. J. Zucker. (1995) "Childhood Sex-Typed Behavior and Sexual Orientation: A Conceptual Analysis and Quantitative Review," *Developmental Psychology* 31: 43-55.

Bailey, J. M. et al. (2000) "Genetic and Environmental Influence on Sexual Orientation and Its Correlates in an Australian Twin Sample," *Journal of Personality and Society Psychology* 78: 524–36.

———. (1993) "Heritable Factors Influence Sexual Orientation in Women," *Archives of General Psychiatry* 50: 217–23.

Bailey, Michael. *The Man Who Would Be Queen: The Science of Gender-Bending and Transsexualism*. Washington, DC: Joseph Henry Press, 2003.

Barber, Nigel. "Gay Animals," *Why We Do What We Do, Psychology Today*, June 17, 2009.

Baroncini, M. et al. (2010) "Sex Steroid Hormones-Related Structural Plasticity in the Human Hypothalamus," *NeuroImage* 50: 428–43.

Bartlett, N. H. et al. (2000) "Is Gender Identity Disorder in Children a Mental Disorder?" *Sex Roles* 43: 753–85.

Beck, Max. "Intersexual Life," *NOVA,* pbs.org, October 30, 2001.

Bell, Vaughan. "Return of the Gay Brain," *Mind Hacks,* June 17, 2008.

Bem, S. L. (1974) "The Measurement of Psychological Androgyny," *Journal of Consulting and Clinical Psychology* 42: 155–62.

Berenbaum, S., and M. Hines. (1992) "Early Androgens are Related to Childhood Sex-Typed Toy Preferences," *Psychological Science* 3: 203–6.

Bishop, K., and D. Wahlsten. (1997) "Sex Differences in the Human Corpus Callosum: Myth or Reality?" *Neuroscience and Biobehavioral Reviews* 21: 581–601.

Bocklandt, S. et al. (2006) "Extreme Skewing of X Chromosome Inactivation in Mothers of Homosexual Men," *Human Genetics* 118: 691–94.

Bornstein, Kate, and S. Bear Bergman. *Gender Outlaws: The Next Generation.* Berkeley, CA: Seal Press, 2010.

Brown, Patricia. "Supporting Boys or Girls When the Line Isn't Clear," *New York Times*, December 2, 2006.

Burri, A. et al. (2011) "Genetic and Environmental Influences on Female Sexual Orientation, Childhood Gender Typicality and Adult Gender Identity," *PLoS One* 6: e21982.

Carey, Benedict. "Psychiatry Giant Sorry for Backing Gay 'Cure,'" *New York Times*, May 18, 2012.

Carothers, B. J., and H. T. Reis. (2012) "Men and Women Are from Earth: Examining the Latent Structure of Gender," *Journal of Personality and Social Psychology* 104: 385–407.

Case, L., and V. S. Ramachandran. (2012) "Alternating Gender Incongruity: A New Neuropsychiatric Syndrome Providing Insight into the Dynamic Plasticity of Brain-Sex," *Medical Hypotheses* 78: 626–31.

Cherney, I. D. and K. London. (2006) "Gender-Linked Differences in the Toys, Television Shows, Computer Games, and Outdoor Activities of 5- to 13-year-old Children," *Sex Roles* 54: 717–26.

Cherney, I. D. et al. (2003) "The Effects of Stereotyped Toys and Gender on Play Assessment in Children Aged 18–47 Months," *Educational Psychology: An International Journal of Experimental Educational Psychology* 23: 95–106.

Chivers, M. L. (2010) "A Brief Update on the Specificity of Sexual Arousal," *Sexual and Relationship Therapy* 25: 407–14.

Chivers, M. L. et al. (2007) "Gender and Sexual Orientation Differences in Sexual Response to the Sexual Activities Versus the Gender of Actors in Sexual Films," *Journal of Personality and Social Psychology* 93: 1108–121.

Cohen-Bendahan, C. C. C. et al. (2005) "Prenatal Sex-Hormone Effects on Child and Adult Sex-Typed Behavior: Methods and Findings," *Neuroscience and Biobehavioral Reviews* 29: 353–84.

Costandi, Mo. "The Science and Ethics of Voluntary Amputation," Neurophilosophy, *The Guardian*, May 30, 2012.

Del Giudice, M. et al. (2012) "The Distance Between Mars and Venus: Measuring Global Sex Differences in Personality," *PLoS One* 7: e29265

DeVries, G. J., and R. B. Simerly. "Anatomy, Development and Function of Sexually Dimorphic Neural Circuits in the Mammalian Brain." In *Hormones, Brain and Behavior*. Donald Pfaff et al., eds. San Diego: Academic Press, 2002.

Diamond, Lisa. *Sexual Fluidity: Understanding Women's Love and Desire*. Cambridge, MA: Harvard University Press, 2009.

Diamond, M. (2002) "Sex and Gender Are Different: Sexual Identity and Gender Identity Are Different," *Clinical Child Psychology and Psychiatry* 7: 320–34.

Diamond, M., and H. K. Sigmundson. (1997) "Sex Assignment at Birth: Long-Term Review and Clinical Implications," *Archives of Pediatric and Adolescent Medicine* 151: 298–394.

Dreger, A. (2009) "Gender Identity Disorder in Childhood: Inconclusive Advice to Parents," *Hastings Center Report* 39: 26–29.

———. (1998) "A History of Intersexuality, from the Age of Gonads to the Age of Consent," *Journal of Clinical Ethics* 9: 345–55.

Drummond, K. D. (2008) "A Follow-Up Study of Girls with Gender Identity Disorder," *Developmental Psychology* 44: 34–45.

DuPree, M. G. et al. (2004) "A Candidate Gene Study of CYP19 (Aromatase) and Male Sexual Orientation," *Behavior Genetics* 34: 243–50.

Dvorak, Petula. "Drug Treatments for Transgender Kids Pose Difficult Choices for Parents, Doctors," *Washington Post*, May 19, 2012.

———. "Transgender at Five," *Washington Post*, May 19, 2012.

Dvorsky, George. "Testing Female Athletes to Make Sure They're Feminine Enough Is Not Cool, Scientists Say," io9.com, June 14, 2012.

Esposito, Lisa. "Gender Identity Issues Can Harm Kids' Mental Health," WJMJ.com, February 20, 2012.

Fausto-Sterling, Anne. *Sexing the Body: Gender Politics and the Construction of Sexuality*. New York: Basic Books, 2000.

———. (2012) "The Dynamic Development of Gender Variability," *Journal of Homosexuality* 59: 398–421.

———. "The Five Sexes: Why Male and Female Are Not Enough," *The Sciences*, March/April 1993, 20–4.

———. (2000) "The Five Sexes, Revisited," *The Sciences* 40: 18–23.

Featherstone, D. E. (2011) "Glial SLC Transporters in Drosophila and Mice," *GLIA* 59: 1351–363.

Featherstone, D. E., and S. A. Shippy. (2008) "Regulation of Synaptic Transmission by Ambient Extracellular Glutamate," *The Neuroscientist* 14: 171–81.

Fine, Cordelia. *Delusions of Gender: How Our Minds, Society and Neurosexism Create a Difference*. New York: W. W. Norton, 2010.

First, M. B. (2005) "Desire for Amputation of a Limb: Paraphilia, Psychosis, or a New Type of Identity Disorder," *Psychological Medicine* 35: 919–28.

Friedman, R. C., and J. I. Downey. (1994) "Homosexuality," *The New England Journal of Medicine* 331: 923–30.

Gardner, Amanda. "Gay Men, Straight Women Have Similar Brains," *Live Science*, June 16, 2008.

Garfield, Simon. "Frock Tactics," *The Guardian*, May 27, 2001.

Gentile, D. A. (1993) "Just What are Sex and Gender Anyway?" *Psychological Science* 4: 120–22.

Giedd, J. N., and J. L. Rapoport. (2010) "Structural MRI of Pediatric Brain Development: What Have We Learned and Where Are We Going?" *Neuron* 67: 728–34.

Giummarra, M. J. et al. (2011) "Body Integrity Disorder: Deranged Body Processing, Right Fronto-Parietal Dysfunction, and Phenomenological Experience of Body Incongruity," *Neuropsychological Review*, November 17, 2011.

Gorman, Anna. "Transgender Kids Get Help Navigating a Difficult Path," *Los Angeles Times*, June 15, 2012.

Gorski, R. A. et al. (1980) "Evidence for the Existence of a Sexually Dimorphic Nucleus in the Preoptic Area of the Rat," *Journal of Computational Neurology* 193: 529–39.

Grindley, Lucas. "Cynthia Nixon: Being Bisexual 'Is Not a Choice,'" *The Advocate*, January 30, 2012.

Groff, Philip, and Laura McRae. "The Nature-Nurture Debate in Thirteenth-Century France," Medievalists.net, April 27, 2010.

Grosjean, Y. et al. (2008) "A Glial Amino Acid Transporter Controls Synapse Strength and Courtship in Drosophila," *Nature Neuroscience* 11: 54–61.

Halpern, D. F. (2010) "How Neuromythologies Support Sex Role Stereotypes," *Science* 330: 1320-1321.

Hamer, Dean, and Peter Copeland. *Living with Our Genes*. New York: Anchor Books, 1999.

———. *The Science of Desire: The Search for the Gay Gene and the Biology of Behavior*. New York: Simon & Schuster, 1994.

Hamer, D. et al. (1993) "A Linkage Between DNA Markers on the X Chromosome and Male Sexual Orientation," *Science* 261: 321–27.

Harris, J. A. et al. (1998) "The Heritability of Testosterone: a Study of Dutch Adolescent Twins and Their Parents," *Behavior Genetics* 28: 165–71.

Harris, Paul, and Andrew Levy. "Boy, 5, Lives Life as Girl: Youngest Diagnosed with Gender Identity Disorder," *The Daily Mail*, February 20, 2012.

Hassett, J. et al. (2008) "Sex Differences in Rhesus Monkey Toy Preferences Parallel Those of Children," *Hormones and Behavior* 54: 359–64.

Hauschka, T. S. et al. (1962) "An XYY Man with Progeny Indicating Familial Tendency to Non-Disjunction," *American Journal of Human Genetics* 14: 22–30.

Herbenick, D. et al. (2010) "Sexual Behavior in the United States: Results From a National Probability Sample of Men and Women Ages 14–94," *Journal of Sexual Medicine* 7: 255–65.

Herman, S. P. (1983) "Gender Identity Disorder in a Five-Year-Old Boy," *The Yale Journal of Biology and Medicine* 56: 15–22.

Hines, Melissa. *Brain Gender*. New York: Oxford University Press, 2004.

———. (2008) "Early Androgen Influences on Human Neural and Behavioural Development," *Early Human Development* 84: 805–07.

———. "Gonadal Hormones and Sexual Differentiation of Human Brain and Behavior." In *Hormones, Brain and Behavior*, 2nd edition, 3: 1869–1909. Edited by Donald Pfaff et al. New York: Academic Press, 2009.

———. "Play and Gender." In *The Child: An Encyclopedic Companion*. Edited by Richard A. Shweder et al. Chicago: University of Chicago Press, Chicago, 2009.

———. (2006) "Prenatal Testosterone and Gender-Related Behavior," *European Journal of Endocrinology* 155: S115–S121.

Hines, M. et al. (2004) "Androgen and Psychosexual Development: Core Gender Identity, Sexual Orientation and Recalled Childhood Gender Role Behavior in Men and Women with Congenital Adrenal Hyperplasia (CAH)," *Journal of Sex Research* 41: 75–81.

———. (2003) "Psychological Outcomes and Gender-Related Development in Complete Androgen Insensitivity Syndrome (CAIS)," *Archives of Sexual Behavior* 32: 93–101.

Hiscock, John. "*Sex and the City's* Cynthia Nixon: 'I'm Just a Woman in Love with a Woman,'" *The Daily Telegraph*, May 13, 2008.

Hu, S. et al. (1995) "Linkage Between Sexual Orientation and Chromosome Xq28 in Males But Not in Females," *Nature Genetics* 11: 248–56.

Hvistendahl, Mara. "Tweaking a Single Gene in Female Mice Has Been Found to Change Their Sexual Preference," *Popular Science*, July 14, 2010.

Jacobs, P. A. et al. (1965) "Aggressive Behavior, Mental Sub-Normality and the XYY Male," *Nature* 208: 1351–352.

Joel, D. (2011) "Male or Female? Brains Are Intersex," *Frontiers in Integrative Neuroscience*, 5: 1–5.

Johnston, Casey. "I Am No Man: For Zelda-Playing Daughter, Dad Gives Link a Sex-Change," *Ars Technica*, November 8, 2012.

Jordan-Young, Rebecca. *Brain Storm: The Flaws in the Science of Sex Differences*. Cambridge, MA: Harvard University Press, 2010.

Kahlenberg, S., and R. Wrangham. (2010) "Sex Differences in Chimpanzees' Use of Sticks as Play Objects Resemble Those of Children," *Current Biology* 20: R1067–R1068.

Kallman, F. J. (1952) "Twin and Sibship Studies of Overt Male Homosexuality," *American Journal of Human Genetics* 4: 136–46.

Kaplan, Karen. "Cynthia Nixon Says She's 'Gay By Choice.' Is It Really a Choice?" *Los Angeles Times*, January 25, 2012.

Karkazis, K. et al. (2012) "Out of Bounds? A Critique of the New Policies on Hyperandrogenism in Elite Female Athletes," *The American Journal of Bioethics* 12: 3–16.

Karr, J. et al. (2009) "Regulation of Glutamate Receptor Availability by MicroRNAs," *Journal of Cell Biology* 185: 685–97.

Khateb, A. et al (2009) "Seeing the Phantom: A Functional MRI Study of a Supernumerary Phantom Limb," *Annals of Neurology* 65: 698–705.

King, Barbara J. "Can Children Know at Age 2 They Were Born the Wrong Sex?" NPR.org, May 24, 2012.

Kinsey, A. et al. *Sexual Behavior in the Human Female*. Philadelphia: Saunders, 1953.

———. *Sexual Behavior in the Human Male*. Philadelphia: Saunders, 1948.

Knickmeyer, R. et al. (2005) "Gender-Typed Play and Amniotic Testosterone," *Developmental Psychobiology* 41: 517–28.

Lefever, Robert. "After a Five-Year-Old Becomes the Youngest Known Child to be Diagnosed with Gender Identity Disorder, Just When Is a Boy a Girl?" *The Daily Mail*, February 21, 2012.

Lerner, Barron H. "If Biology Is Destiny, When Shouldn't It Be?" *The New York Times*, May 27, 2003.

LeVay, Simon. *Gay, Straight and the Reason Why: The Science of Sexual Orientation*. New York: Oxford University Press, 2011.

———. *The Sexual Brain*. Cambridge: MIT Press, 1993.

———. (1991) "A Difference in Hypothalamic Structure Between Heterosexual and Homosexual Men," *Science* 253: 1034–37.

———. "The Gay Brain Revisited," *Nerve*, September 5, 2000.

LeVay, Simon, and Dean Hamer. "Evidence for a Biological Influence in Male Homosexuality," *Scientific American*, May 1994.

Lister, David. "It's Never Too Late in Edinburgh," *The Independent*, August 26, 1992.

Lucas, M. (2012) "Book Review: Cordelia Fine, *Delusions of Gender: How our Minds, Society, and Neurosexism Create Difference*," *Sociology* 49: 199–202.

Luders, E. et al. (2009) "Regional Gray Matter Variation in Male-to-Female Transsexualism," *Neuroimage* 46: 904–7.

Malory, Marcia. "Is Homosexuality a Choice?" Guest Blog, *Scientific American*, October 19, 2012.

Marsa, Linda. "He Thinks, She Thinks," *Discover*, July 2007.

Maugh, Thomas H. II. "Genetic Component Found in Lesbianism, Study Says: Research on Twins Shows that Environment Also Plays a Role, Although Unclear, in Sexual Orientation," *Los Angeles Times*, March 12, 1993.

McCarthy, M. M., and G. F. Ball. (2011) "Tempests and Tales: Challenges to the Study of Sex Differences in the Brain," *Biological Sex Differences* 2: 4.

McCarthy, M. M. et al. (2012) "Sex Differences in the Brain: The Not So Inconvenient Truth," *The Journal of Neuroscience* 32: 2241–247.

McGeoch, P. D. et al. (2008) "Apotemnophilia—the Neurological Basis of a 'Psychological' Disorder," *Neuroreport*, 19: 1305–6.

———. (2011) "Xenomelia: a New Right Parietal Lobe Syndrome," *Neurological and Neurosurgical Psychiatry* 82: 1314–19.

McKie, Robin. "'Sexual Depravity' of Penguins that Antarctic Scientist Dared Not Reveal," *The Guardian*, June 9, 2012.

McKissick, Katie. "Biology Doesn't Support Gay Marriage Bans," *Beatrice the Biologist*, February 8, 2012.

———. "Clarification: Sex Determination and Gay Marriage," *Beatrice the Biologist*, February 13, 2012.

Mooallem, Jon. "Can Animals Be Gay?" *New York Times*, March 31, 2010.

Mustanski, B. S. et al. (2005) "A Genome-Wide Scan of Male Sexual Orientation," *Human Genetics* 116: 272–78.

Ohlheiser, Abby. "How Many Americans Self-Identify as Gay, Lesbian, Bisexual, or Transgender?" *Slate*, October 19, 2012.

Owen, James. "Gay Men, Straight Women Have Similar Brains," *National Geographic Daily News*, June 16, 2008.

Owen-Blakemore, J. E., and R. E. Centers. (2005) "Characteristics of Boys' and Girls' Toys," *Sex Roles* 53: 619–33.

Padawer, Ruth. "What's So Bad About a Boy Who Wants to Wear a Dress?" *New York Times*, August 8, 2012.

Pahle, Rebecca. "13-Year-Old Girl Whose Little Brother Wants an Easy-Bake Oven for Christmas Petitions Hasbro for a Gender-Neutral Version," *The Mary Sue*, December 3, 2012.

Park, Alice. "What the Gay Brain Looks Like," *Time*, June 17, 2008.

Parma, P. et al. (2006) "R-spondin1 is Essential in Sex Determination, Skin Differentiation and Malignancy," *Nature Genetics* 38: 1304–309.

Pasterski, V. L. et al. (2007) "Increased Aggression and Activity Level in 3- to 11-Year-Old Girls with Congenital Adrenal Hyperplasia," *Hormones and Behavior* 52: 368–74.

———. (2005) "Prenatal Hormones Versus Postnatal Socialization by Parents as Determinants of Male-Typical Toy Play in Girls with Congenital Adrenal Hyperplasia," *Child Development* 76: 264–78.

Pattatucci, A., and D. Hamer. (1995) "Development and Familiarity of Sexual Orientation in Females," *Behavior Genetics* 25: 407–19.

Pfaff, Donald. *Man and Woman: An Inside Story*. Oxford, UK: Oxford University Press, 2010.

Pickstone-Taylor, S. D. (2003) "Children with Gender Nonconformity," *Journal of the American Academy of Child and Adolescent Psychiatry* 42: 266.

Prinz, Jesse. *Beyond Human Nature: How Culture and Experience Shape Our Lives*. New York: Penguin, 2012.

Pryzgoda, J., and J. C. Chrisler. (2000) "Definitions of Gender and Sex: The Subtleties of Meaning," *Sex Roles* 43: 553–69.

Ramachandran, V. S., and W. Hirstein. (1998) "The Perception of Phantom Limbs: The D. O. Hebb Lecture," *Brain* 121:1603–630.

Ramachandran, V. S., and P. D. McGeoch. (2008) "Phantom Penises in Transsexuals: Evidence of an Innate Gender-Specific Body Image in the Brain," *Journal of Consciousness Studies* 15: 5–16.

Rametti, G. et al. (2011) "The Microstructure of White Matter in Male to Female Transsexuals Before Cross-Sex Hormonal Treatment. A DTI Study," *Journal of Psychiatric Research* 45: 949–54.

Rampton, James. "Eddie Izzard: The Tough Transvestite Who Can Take Care of Himself," *The Independent*, May 23, 2004.

Reed, Natalie. "Bilaterally Gynandromorphic Chickens and Why I'm Not Scientifically Male," *Sincerely, Natalie Reed*, Freethought Blogs, March 28, 2012.

————. "Gender Expression Is Not Gender Identity," *Sincerely, Natalie Reed,* Freethought Blogs, March 21, 2012.

————. "Myths and Misconceptions About Trans Women," *Sincerely, Natalie Reed,* Freethought Blogs, March 27, 2012.

Rice, G. et al. (1999). "Male Homosexuality: Absence of Linkage to Microsatellite Markers at Xq28," *Science* 284: 665–67.

Rice, W. R. et al. (2012) "Homosexuality as a Consequence of Epigenetically Canalized Sexual Development," *The Quarterly Review of Biology* 87: 1–25.

Rieger, G. et al. (2005) "Sexual Arousal Patterns of Bisexual Men," *Psychological Science* 16: 579–84.

————. (2008) "Sexual Orientation and Childhood Gender Nonconformity: Evidence from Home Videos," *Developmental Psychology* 44: 46–58.

Roselli, C. E. et al. (2008) "Prolactin Expression in the Sheep Brain," *Neuroendocrinology* 87: 206–15.

Rosin, Hannah. "A Boy's Life," *The Atlantic*, November 2008.

Roughgarden, Joan. *Evolution's Rainbow: Diversity, Gender and Sexuality in Nature and People*. Berkeley, CA: University of California Press, 2004, 2009.

Rubin, Gayle S. "Thinking Sex: Notes for a Radical Theory of the Politics of Sexuality." In *Pleasure and Danger*. Edited by Carole Vance. London, UK: Pandora, 1992, 267–93.

Saletan, William. "Sex on the Brain: Are Boys Brains Different from Girls' Brains? Yes and No," *Slate*, November 17, 2011.

Savic-Bergland, I., and P. Lindstrom. (2008) "PET and MRI Show Differences in Cerebral Asymmetry and Functional Connectivity Between Homo- and Heterosexual Subjects," *Proceedings of the National Academy of Science* 105: 9403–408.

Schaffer, Amanda. "The Last Word on Fetal T," *Slate*, October 21, 2010.

Spiegel, Alix. "Two Families Grapple with Sons' Gender Preferences," *All Things Considered*, NPR, May 7, 2008.

Stix, Gary. "Ramachandran Lab Looks into Whether You Can Be a Man in the Morning and a Woman at Night," *Observations, Scientific American*, April 19, 2012.

Suschinsky, K. et al. (2009) "Sex Differences in Patterns of Genital Arousal: Measurement Artifact or True Phenomenon?" *Archives of Sexual Behavior* 38: 559–73.

Swaab, D. F. (2004) "Sexual Differentiation of the Human Brain: Relevance for Gender Identity, Transsexualism, and Sexual Orientation," *Gynecological Endocrinology* 19: 301–12.

Swidey, Neil. "What Makes People Gay," *Boston Globe*, August 14, 2005.

Tennent, W. J. (1987) "A Note on the Apparent Lowering of Moral Standards in the Lepidoptera," *Entomologist's Record and Journal of Variation* 99: 81–83.

Udry, J. R. et al. (1995) "Androgen Effects on Women's Gendered Behavior," *Journal of Biosocial Science* 27: 359–68.

Walker, Jesse. "The Death of David Reimer," Reason.com, May 24, 2004.

Weinstein, N. et al. (2012) "Parental Autonomy Support and Discrepancies Between Implicit and Explicit Sexual Identities: Dynamics of Self-Acceptance and Defense," *Journal of Personality and Social Psychology* 102: 815–32.

Whitman, F. L. et al. (1993) "Homosexual Orientation in Twins: a Report on 61 Pairs and Three Triplet Sets," *Archives of Sexual Behavior* 22: 187–206.

Wickelgren, I. (1999). "Discovery of 'Gay Gene' Questioned," *Science* 284: 571.

Wood, J. L. et al. (2008) "Morphology of the Ventral Frontal Cortex: Relationship to Femininity and Social Cognition," *Cerebral Cortex* 18: 534–40.

Xu, X. et al. (2012) "Modular Genetic Control of Sexually Dimorphic Behaviors," *Cell* 148: 596–607.

Zhang, S., and W. Odenwald. (1995) "Misexpression of the White (W) Gene Triggers Male-Male Courtship in Drosophila," *Proceedings of the National Academy of Sciences* 92: 5525–529.

Zucker, K. J. (2008) "Children with Gender Identity Disorder: Is There a Best Practice?" *Neuropsychiatrie de l'Enfance et de l'Adolescence* 56: 358–64.

Online Resources

Center for Gender Sanity: http://www.gendersanity.com

Gallup Report on LGBT Self-Identification: http://www.gallup.com/poll/158066/special-report-adults-identify-lgbt.aspx

David Featherstone's fly courtship movies: http://tigger.uic.edu/~def/gb/Featherstone_lab_courtship_movies/Featherstone_Lab_fly_courtship_movies.html

The Colbert Report on "albatresbians": http://www.colbertnation.com/the -colbert-report-videos/270727/april-13-2010/jon-mooallem

Alice Dreger TED Talk: "Is Anatomy Destiny?" http://www.ted.com/talks/ alice_dreger_is_anatomy_destiny.html

Interview with Lisa Diamond: http://www.hup.harvard.edu/news/audio/ DIASEX.mp3

Transgender basics video: http://www.gaycenter.org/transgenderbasics

The Lesbian, Gay, Bisexual and Transgender Community Center: http:// www.gaycenter.org/

The Bisexuality Report: http://www8.open.ac.uk/ccig/files/ccig/The%20 BisexualityReport%20Feb.2012_0.pdf

World Science Festival: "The Origins of Orientation: Sexuality in 2011" http://worldsciencefestival.com/videos/the_origins_of_orientation_ sexuality_in_2011

UCSF video on gender-specific behavior in mice: http://www.youtube.com/ watchzv=fsoXZ9_Hh7Y

UCSF News in Gender-Specific Behavior in Mice: http://www.ucsf.edu/ news/2012/02/11440/male-and-female-behavior-deconstructed

III. WHY

7. Feed Your Head

Allen, P. et al. (2008) "The Hallucinating Brain: A Review of Structural and Functional Neuroimaging Studies of Hallucinations," *Neuroscience and Biobehavioral Reviews* 32: 175–91.

Appela, J. B. et al. (2004) "Review: LSD, 5-HT (Serotonin), and the Evolution of a Behavioral Assay," *Neuroscience and Biobehavioral Reviews* 27: 693–701.

Baggott, M. J. et al. (2010) "Investigating the Mechanisms of Hallucinogen-Induced Visions Using 3,4-methylenedioxyamphetamine (MDA): A Randomized Controlled Trial in Humans," *PLoS One* 5: e14074.

Beauchamp, Cari, and Judy Bachrach. "Cary in the Sky with Diamonds," *Vanity Fair*, August 2010.

Bressloff, P. C. et al. (2002) "What Geometric Visual Hallucinations Tell Us about the Visual Cortex," *Neural Computation* 14: 473–91.

Buckholtz, N. S. et al. (1990) "Lysergic Acid Diethylamide (LSD) Adminis-
tration Selectively Downregulates Serotonin2 Receptors in Rat Brain,"
Neuropsychopharmacology 3: 137–48.

Bushwick, Sophie. "Why Don't We Normally Hallucinate?" io9.com, December 30, 2011.

Butler, T. C. et al. (2011) "Evolutionary Constraints on Visual Cortex Architecture from the Dynamics of Hallucinations," *Proceedings of the National Academy of Sciences,* 109: 606–609.

Carhart-Harris, R. L., and K. J. Friston. (2010) "The Default-Mode, Ego-
Functions and Free-Energy: A Neurobiological Account of Freudian
Ideas," *Brain* 133: 1265–283.

Carhart-Harris, R. L. et al. (2012) "Neural Correlates of the Psychedelic State
as Determined by fMRI Studies with Psilocybin," *Proceedings of the National Academy of Sciences,* 109: 2138–2143.

Carroll, Sean M. "Hallucinatory Neurophysics," *Preposterous Universe,* February 4, 2005.

Chwelos, N. et al. (1959) "Use of D-Lysergic Acid Diethylamide in the Treatment of Alcoholism," *Quarterly Journal of Studies in Alcoholism* 20:
577–90.

Costandi, Mo. "Magic Mushrooms in the Neuropsychoanalytical Framework," *Neurophilosophy, The Guardian,* January 25, 2012.

———. "Psychedelic Chemical Subdues Brain Activity," Nature.com, January 23, 2012.

———. "Seeing with Eyes Wide Shut: Ayahuasca Inner Vision," *Neurophilosophy, The Guardian,* September 23, 2011.

Dalí, Salvador. *Dalí by Dalí.* New York: Harry N. Abrams, 1970.

de Araujo, D. B. et al. (2011) "Seeing with the Eyes Shut: Neural Basis of Enhanced Imagery Following Ayahuasca Ingestion," *Human Brain Mapping,* September 16, 2011.

de Rios, Marlene Dobkin, and Oscar Janiger. *LSD, Spirituality, and the Creative Process.* Rochester, VT: Inner Traditions, 2003.

Ferguson, Craig. *American on Purpose: The Improbable Adventures of an
Unlikely Patriot.* New York: Harper, 2009.

Friston, Karl. "The Free-Energy Principle: A Unified Brain Theory?" *Nature
Reviews Neuroscience,* January 13, 2010.

Gonzalez, Robert. "Ten Scientific and Technological Visionaries Who Experimented with Drugs," io9.com, January 16, 2012.

Greenfield, Robert. *Timothy Leary: A Biography*. New York: Harcourt Books, 2006.

Griffiths, R. R. et al. (2008). "Mystical-Type Experiences Occasioned by Psilocybin Mediate the Attribution of Personal Meaning and Spiritual Significance 14 Months Later," *Journal of Psychopharmacology* 22: 621–32.

———. (2006) "Psilocybin Can Occasion Mystical Experiences Having Substantial and Sustained Personal Meaning and Spiritual Significance," *Psychopharmacology* 187: 268–83.

Grosse Pointe Blank. (1997) Hollywood Pictures. Dir. George Armitage. Story: Tom Jankiewicz. Screenplay: Tom Jankiewicz, D. V. DeVincentis, Steve Pink, John Cusack.

Halpern, J. H. (1996) "The Use of Hallucinogens in the Treatment of Addiction," *Addiction Research* 4: 177–89.

Halpern, J. H., and H. G. Pope Jr. (2003) "Hallucinogen Persisting Perception Disorder: What Do We Know After 50 Years?" *Drug and Alcohol Dependence* 69: 109–19.

Halpern, J. H. et al. (2011) "Residual Neurocognitive Features of Long-Term Ecstasy Users with Minimal Exposure to Other Drugs," *Addiction* 106: 777–86.

Harrison, Ann. "LSD: The Geek's Wonder Drug," *Wired*, January 16, 2006.

Healy, Melissa. "Turn on, Tune In, and Get Better?" *Los Angeles Times*, November 30, 2011.

Hoffmann, Albert. *LSD: My Problem Child*. San Francisco: Multidisciplinary Association for Psychedelic Studies, 2009.

Horgan, John. "Peyote on the Brain," *Discover*, February 1, 2003.

Huxley, Aldous. *The Doors of Perception*. New York: Harper Perennial, 2009.

Huxley, Laura. *This Timeless Moment: A Personal View of Aldous Huxley*. Berkeley: Celestial Arts, 2000.

Janiger, O., and M. D. de Rios. (1989) "LSD and Creativity," *Journal of Psychoactive Drugs* 21: 129–34.

Keim, Brandon. "Alan Turing's Patterns in Nature, and Beyond," Wired Science, *Wired*, February 22, 2011.

———. "Psilocybin Study Hints at Rebirth of Hallucinogen Research," *Wired Science, Wired,* July 1, 2008.

Koch, Christof. "This Is Your Brain on Drugs," *Scientific American Mind*, May/June 2012.

Kornell, Sam. "Researchers Re-Open Their Minds to Psychedelic Drugs," *Miller-McCune*, May 5, 2011.

Lattin, Don. *The Harvard Psychedelic Club: How Timothy Leary, Ram Dass, Huston Smith, and Andrew Weil Killed the Fifties and Ushered in a New Age for America.* New York: HarperCollins, 2011.

———. "Leary's Legacy," *UC-Berkeley Alumni Magazine*, Fall 2010.

Leary, Timothy. *The Psychedelic Experience: A Manual Based on the Tibetan Book of the Dead.* Berkeley, CA: Citadel Underground, 2000.

———. *Turn on, Tune In, Drop Out.* Berkeley, CA: Ronin Publishing, 1999.

Lewin, L. (1894) "On *Anhalonium lewinii* and Other Cacti," *Archiv für Experimentelle Pathologie und Pharmakologie.*

Mash, D. C. et al. (1998) "Medication Development of Ibogaine as a Pharmacotherapy for Drug Dependence." In *The Neurology of Drugs of Abuse: Cocaine, Ibogaine and Substituted Amphetamines.* Edited by Syed Ali. Annals of the New York Academy of Science, 275–92.

Meltzer, Marisa. "Shamanism and the City," *GOOD*, November 22, 2011.

McKie, Robin. "Ecstasy Does Not Wreck the Mind, Study Claims," *The Guardian*, February 19, 2011.

Moore, Galen. "Harvard's Headache Cure: LSD?" *Mass High Tech,* November 5, 2010.

Naditch, M. P., and S. Fenwick. (1977) "LSD Flashbacks and Ego Functioning," *Journal of Abnormal Psychology* 86: 352–59.

Nichols, David E. "LSD: Cultural Revolution and Medical Advances," *Chemistry World*, January 2006.

———. (2004) "Psychotropics," *Pharmacology and Therapeutics* 101: 131–81.

Ouellette, Jennifer. "Biologists Home in on Turing Patterns," *Quanta,* March 2013.

———. "A Long Strange Trip," *Mental Floss*, January/February 2013.

Pappas, Stephanie. "Trippy Tales: The History of 8 Hallucinogens," *Live Science,* September 29, 2011.

Passie, T. et al. (2008) "The Pharmacology of Lysergic Acid Diethylamide: A Review," *CNS Neuroscience & Therapeutics* 14: 295–314

Sanchez-Ramos, J. (1991) "Banisterine and Parkinson's Disease," *Clinical Neuropharmacology* 14: 391–402.

Seltzer, Sarah. "Five Fascinating New Uses for Psychedelics," AlterNet.com May 1, 2012.

Serrano-Duenas, M. et al. (2001) "Effects of *Bainsteriopsis caapi* Extract on Parkinson's Disease," *The Scientific Review of Alternative Medicine* 5: 127–32.

Sessa, B. (2008) "Is It Time to Revisit the Role of Psychedelic Drugs in Enhancing Human Creativity?" *Journal of Psychopharmacology* 22: 821–27.

Sewell, R. A. et al. (2006) "Response of Cluster Headache to Psilocybin and LSD," *Neurology* 66: 1920–922.

Slater, Lauren. "How Psychedelic Drugs Can Help Patients Face Death," *The New York Times*, April 20, 2012.

Stevens, Jay. *Storming Heaven*. New York: Grove Press, 1987.

Szalavitz, Maia. "Magic Mushrooms Can Improve Psychological Health Long Term," *Time*, June 16, 2011.

———. "Magic Mushrooms Expand the Mind by Dampening Brain Activity," *Time*, January 24, 2012.

Tierney, John. "Hallucinogens Have Doctors Tuning In Again," *The New York Times*, April 12, 2010.

Turing, A. (1952) "The Chemical Basis of Morphogenesis," *Philosophical Transactions of the Royal Society of London. Series B, Biological Sciences* 237: 37–72.

Unger, S. M. (1963) "Mescaline, LSD, Psilocybin and Personality Change," *Psychiatry: Journal for the Study of Interpersonal Processes* 26(2).

Usher, Shaun. "The Most Beautiful Death," *Letters of Note*, March 25, 2010.

Weil, Andrew T. "The Strange Case of the Harvard Drug Scandal," *Look*, November 5, 1963.

Whelan, John. "The Trip: Cary Grant on Acid, and Other Stories from the LSD Studies of Dr. Oscar Janiger," *LA Weekly*, July 9, 1998.

Wilde, Alex. "The Acid Test," *Cosmos*, November 7, 2007.

Online Resources

Kiki Sanford interviews Matthew Baggott on psychedelics: http://www .youtube.com/watch?v=XxMZg1oEWd4

Robin Carhart-Harris 2010 talk at MAPS (video): http://www.maps.org/ videos/source2/video6.html

Juan Sanchez-Ramos 2010 MAPS talk part 1: http://www.youtube.com/
    watch?v=eV3l1YIpdik

Juan Sanchez-Ramos 2010 MAPS talk part 2: http://www.youtube.com/
    watch?v=PD17vETLC-I

Juan Sanchez-Ramos lecture on Vimeo: http://vimeo.com/33812651

8.   Ghost in the Machine

Alkire, M. T. et al. (2008) "Consciousness and Anesthesia," *Science* 322:
    876–80.

Anderson, P. W. (1972) "More Is Different: Broken Symmetry and the Nature
    of the Hierarchical Structure of Science," *Science* 177: 393–96.

Bell, Vaughan. "Anesthesia May Leave Patients Conscious—and Finally
    Show Consciousness in the Brain," *The Crux, Discover*, January 4, 2012.

Bendor, D., and M. A. Wilson. (2012) "Biasing the Content of Hippocampal
    Replay During Sleep," *Nature Neuroscience* 15: 1439–444.

Bonabeau, E. (2002) "Predicting the Unpredictable," *Harvard Business
    Review* 80: 109–16.

Bor, Daniel. *The Ravenous Brain: How the New Science of Consciousness
    Explains Our Insatiable Search for Meaning.* New York: Basic Books,
    2012.

Brown, E. N. et al. (2010) "General Anesthesia, Sleep, and Coma," *New En-
    gland Journal of Medicine* 363: 2638–650.

Buldyrev, S. et al. (2010) "Catastrophic Cascade of Failures in Interdependent
    Networks," *Nature* 464: 1025–28.

Bullmore, E., and O. Sporns. (2012) "The Economy of Brain Network Orga-
    nization, *Nature Reviews Neuroscience* 13: 336–49.

Castro, Jason. "The Brain May Disassemble Itself in Sleep," *Scientific Amer-
    ican*, May 24, 2012.

Chalmers, D. (1995) "Facing Up to the Problem of Consciousness," *Journal of
    Consciousness Studies* 2: 200–19.

———. "What Is a Neural Correlate of Consciousness?" In *Neural Corre-
    lates of Consciousness: Empirical and Conceptual Questions.* Edited by
    Thomas Metzinger. Cambridge: MIT Press, 2000, 17–40.

Churchland, Patricia. *The Engine of Reason, The Seat of the Soul: A Philo-
    sophical Journey into the Brain.* Cambridge, MA: MIT Press, 1995.

———. "How Do Neurons Know?" *Daedalus*, Winter 2004.

———. (2002) "Self-Representation in Nervous Systems," *Science* 296: 308–10.

Cohen, Philip. "Small World Networks Key to Memory," *New Scientist*, 26 May 2004.

Corkin, S. (2002) "What's New with the Amnesic Patient H. M.?" *Nature Reviews in Neuroscience* 3: 153–60.

Corning, P. A. (2002) "The Re-Emergence of 'Emergence': A Venerable Concept in Search of a Theory," *Complexity* 7: 18–30.

Costandi, Mo. "Remembering Henry M.," *Neurophilosophy*, July 28, 2007.

Crick, Francis. *The Astonishing Hypothesis: The Scientific Search for the Soul*. New York: Scribner, 1995.

Crick, F., and C. Koch. (2002) "A Framework for Consciousness," *Nature Neuroscience* 6: 119–26.

———. (1990) "Towards a Neurobiological Theory of Consciousness," *Seminars in the Neurosciences* 2: 263–75.

Cyranoski, David. "Do Brain Scans of Comatose Patients Reveal a Conscious State?" *Nature*, June 13, 2012.

Damasio, Antonio. *Descartes' Error*. New York: Grosset/Putman, 1994.

———. *Self Comes to Mind: Constructing the Conscious Brain*. New York: Pantheon Books, 2011.

Damasio, A. et al. (1985) "Amnesia Following Basal Forebrain Lesions," *Archives of Neurology* 42: 263–71.

———. (1985) "Multimodal Amnesic Syndrome Following Bilateral Temporal and Basal Forebrain Damage," *Archives of Neurology* 42: 252–59.

Davidson, T. J. et al. (2009) "Hippocampal Replay of Extended Experience," *Neuron* 63: 497–507.

Dehaene, S., and L. Naccache. (2001) "Towards a Cognitive Neuroscience of Consciousness: Basic Evidence and a Workspace Framework," *Cognition* 79: 1–37.

Dehaene, S. et al. (2001) "Cerebral Mechanisms of Word Masking and Unconscious Repetition Priming," *Nature Neuroscience* 4: 752–58.

Dennett, Daniel C. *Consciousness Explained*. Boston: Little Brown, 1991.

Descartes, René. *The Passions of the Soul*. Translated by Steven H. Voss. New York: Hackett Publishing, 1989.

Dvorsky, George. "Rare Neurological Patient Shows that Self-Awareness Does Not Require a Complex Brain," io9.com, August 23, 2012.

Fass, Craig et al. *Six Degrees of Kevin Bacon*. New York: Plume, 1996.

Faugeras, F. et al. (2011) "Probing Consciousness with Event-Related Potentials in Patients Who Meet Clinical Criteria for Vegetative State," *Neurology* 77: 264–68.

Feinberg, Todd E. *Altered Egos: How the Brain Creates the Self*. Oxford, UK: Oxford University Press, 2001.

Geddes, Linda. "Banishing Consciousness: The Mystery of Anesthesia," *New Scientist*, November 29, 2011.

Goldfine, A. M., and N. D. Schiff. (2011) "Consciousness: Its Neurobiology and the Major Classes of Impairment," *Neurologic Clinics* 29: 723–37.

Goldfine, A. M. et al. (2012) "Bedside Detection of Awareness in the Vegetative State," *The Lancet* 379: 1701–702.

Goldstein, J. (1999) "Emergence as a Construct: History and Issues," *Emergence: Complexity and Organization* 1: 49–72

Gorman, James. "Awake or Knocked Out? The Line Gets Blurrier," *The New York Times*, April 12, 2012.

Harman, Katherine. "Octopuses Gain Consciousness (According to Scientists' Declaration)," *Octopus Chronicles, Scientific American*, August 21, 2012.

He, Y. et al. (2007) "Small-World Anatomical Networks in the Human Brain Revealed by Cortical Thickness from MRI," *Cerebral Cortex* 17: 2407–419.

Hilgetag, C. C. et al. (2000) "Anatomical Connectivity Defines the Organization of Clusters of Cortical Areas in the Macaque and the Cat," *Philosophical Transactions of the Royal Society of London B, Biological Sciences* 355: 91.

Humphries, Courtney. "The Mystery Behind Anesthesia," *Technology Review*, January/February 2012.

Huxley, Julian S., and Thomas Henry Huxley. *Evolution and Ethics: 1893–1943*. London: The Pilot Press, 1947.

Interlandi, Jeneen. "A Drug That Wakes the Near Dead," *The New York Times*, December 1, 2011.

Jabr, Ferris. "Does Self-Awareness Require a Complex Brain?" *Brain Waves, Scientific American*, August 22, 2012.

————. "Me, Myself and My Stranger: Understanding the Neuroscience of Selfhood," *Scientific American*, September 21, 2010.

Johnson, Steven. *Emergence: The Connected Lives of Ants, Brains, Cities, and Software*. New York: Scribner, 2001.

Koch, Christof. *Consciousness: Confessions of a Romantic Reductionist*. Cambridge, MA: MIT Press, 2012.

————. *The Quest for Consciousness: A Neurobiological Approach*. New York: Roberts and Co., 2004.

————. "Consciousness Is Everywhere," *Huffington Post*, August 15, 2012.

————. "Safely Switching Consciousness Off and On Again," *Scientific American*, September 20, 2012.

Koestler, Arthur. *The Ghost in the Machine*. New York: Penguin, 1967 (1990 reprint edition).

Lang, Joshua. "Awakening," *The Atlantic*, January 2013.

Langsjo, J. et al. (2012) "Returning from Oblivion: Imaging the Neural Core of Consciousness," *The Journal of Neuroscience* 32: 4935–943.

LeDoux, Joseph. *The Emotional Brain: The Mysterious Underpinnings of Emotional Life*. New York: Simon & Schuster, 1996.

————. *The Synaptic Self: How Our Brains Become Who We Are*. New York: Penguin, 2003.

Leicht, E. A., and R. M. D'Souza. (2009) "Percolation on Interacting Networks" (preprint), July 6, 2009. http://arxiv.org/pdf/0907.089401.pdf.

Levitin, Daniel. "Amnesia and the Self That Remains When Memory Is Lost," *The Atlantic*, December 2012.

Lewes, G. H. *Problems of Life and Mind* (First Series), 2. London: Trübner, 1875.

Lewis, L. D. et al. (2012) "Rapid Fragmentation of Neuronal Networks at the Onset of Propofol-Induced Unconsciousness," *Proceedings of the National Academy of Sciences* 109: E3377–E3386.

Llinas, Rodolfo R. *I of the Vortex: From Neurons to Self*. Cambridge, MA: MIT Press, 2002.

Locke, John. *An Essay Concerning Human Understanding*. New York: Penguin Books, 1997.

MacDougall, Duncan. "The Soul: Hypothesis Concerning Soul Substance Together with Experimental Evidence of the Existence of Such Substance," *American Medicine*, April 1907.

MacFarquhar, Larissa. "Two Heads," *The New Yorker*, February 12, 2007.

Markiv, N. T. et al. (2010) "Weight Consistency Specifies Regularities of Macaque Cortical Networks," *Cerebral Cortex* 21: 1254–272.

Masuda, N., and K. Aihara. (2004) "Global and Local Synchrony of Coupled Neurons in Small-World Networks," *Biological Cybernetics* 90: 302–9.

Merker, B. (2007) "Consciousness Without a Cerebral Cortex: A Challenge for Neuroscience and Medicine," *Behavioral and Brain Sciences* 30: 63–81.

Milgram, S. (1967) "The Small World Problem," *Psychology Today* 1: 60–7.

Milner, B. et al. (1968) "Further Analysis of the Hippocampal Amnesic Syndrome: 14-Year Follow-Up Study of H.M.," *Neuropsychologia* 6: 215–34.

Mitchell, Kevin. "The Genetics of Emergent Phenotypes," *Wiring the Brain*, March 21, 2013.

Ouellette, Jennifer. "A Moment When Animals Started to Seem More Like People," *Nautilus*, May 9, 2013.

———. "It's a Small World After All," *Cocktail Party Physics*, *Scientific American*, June 28, 2012.

Owen, A. M. et al. (2009) "A New Era of Coma and Consciousness Science," *Progress in Brain Research* 117: 399–411.

Palca, Joe. "Insights from Broken Brains," *Science*, May 18, 1990.

Parvizi, J., and A. R. Damasio. (2001) "Consciousness and the Brainstem," *Cognition* 79: 135–60.

Philippi, C. L. et al. (2012) "Preserved Self-Awareness Following Extensive Bilateral Brain Damage to the Insula, Anterior Cingulate, and Medial Prefrontal Cortices," *PLoS One* 7: e38413.

Quill, Elizabeth. "When Networks Network," *Science News*, September 22, 2012.

Ramachandran, V. S. *A Brief Tour of Human Consciousness*. New York: PI Press, 2004.

Roach, Mary. *Stiff: The Curious Lives of Human Cadavers*. New York: W. W. Norton, 2003.

Roethke, Theodore. "The Waking." In *The Collected Poems of Theodore Roethke*. New York: Doubleday, 1961.

Sacks, Oliver. "Seeing God in the Third Millennium," *The Atlantic*, December 2012.

Sanders, Laura. "Consciousness Emerges," *Science News*, February 25, 2012.

————. "Emblems of Awareness," *Science News*, February 11, 2012.

Scannell, J. W. et al. (1995) "Analysis of Connectivity in the Cat Cerebral Cortex," *Cerebral Cortex* 15: 1463–483.

————. (1999) "The Connectional Organization of the Cortico-Thalamic System of the Cat," *Cerebral Cortex* 9: 277–99.

Schiff, N. D. "The Neurology of Impaired Consciousness: Challenges for Cognitive Neuroscience." In *The Cognitive Neurosciences* (3rd ed.). Edited by Michael S. Gazzaniga. Cambridge, MA: MIT Press, 2004, 1121–129.

Schiff, N. D. and J. B. Posner. (2007) "Another 'Awakenings,'" *Annals of Neurology* 62: 5–7.

Schiff, N. D., et al. (2007) "Behavioural Improvements with Thalamic Stimulation after Severe Traumatic Brain Injury," *Nature* 448: 600–03.

Scoville, W. B., and B. Milner. (1957) "Loss of Recent Memory after Bilateral Hippocampal Lesions," *Journal of Neurology, Neurosurgery and Psychiatry* 20: 11–21.

Sergent, C. et al. (2005) "Timing of the Brain Events Underlying Access to Consciousness During the Attentional Blink," *Nature Neuroscience* 8: 1285–286.

Siegfried, Tom. "Self as Symbol," *Science News*, February 11, 2012.

Simard, D. et al. (2005) "Fastest Learning in Small-World Neural Networks," *Physics Letters A* 336: 8–15.

Smith-Bassett, D., and E. Bullmore. (2006) "Small-World Brain Networks," *Neuroscientist* 12: 512–23.

Sporns, O. et al. (2004) "Organization, Development and Function of Complex Brain Networks," *Trends in Cognitive Science* 8: 418–25.

Squire, Larry R., and Eric R. Kandel. *Memory: From Mind to Molecules*. New York: Freeman, 1999.

Stam, C. J. et al. (2007) "Small-World Networks and Functional Connectivity in Alzheimer's Disease," *Cerebral Cortex* 17: 92–9.

Strogatz, Steven. *Sync: How Order Emerges from Chaos in the University, Nature, and Daily Life*. New York: Hyperion, 2003.

Suhler, C. L., and P. S. Churchland. (2009) "Control: Conscious and Otherwise," *Trends in Cognitive Sciences* 13: 341–47.

Sullivan, John Jeremiah. "One of Us," *Lapham's Quarterly*, Spring 2013.

Tononi, G. (2008) "Consciousness as Integrated Information: A Provisional Manifesto," *Biological Bulletin* 215: 216–42.

Tranel, D., and A. Damasio. (1985) "Knowledge Without Awareness: An Autonomic Index of Facial Recognition by Prosopagnosics," *Science* 228: 1453–454.

Turk, D. J. et al. (2003) "Out of Contact, Out of Mind: The Distributed Nature of Self," *Annals of the New York Academy of Sciences* 1001: 65–78.

Vox, Ford. "State of Mind," *Slate*, April 23, 2009.

Watts, Duncan J. *Small Worlds: The Dynamics of Networks Between Order and Randomness*. Princeton: Princeton University Press, 1999.

Watts, D. J., and S. H. Strogatz. (1998) "Collective Dynamics of 'Small-World' Networks," *Nature* 393: 440–42.

Wyart, V., et al. (2012) "Early Dissociation between Neural Signatures of Endogenous Spatial Attention and Perceptual Awareness during Visual Masking," *Frontiers in Human Neuroscience* 6(16).

Zimmer, Carl. "Sizing Up Consciousness By Its Bits," *The New York Times*, September 20, 2012.

Online Resources

Antonio Damasio TED Talk: http://www.ted.com/talks/antonio_damasio_the_quest_to_understand_consciousness.html

Ned Block on consciousness: http://vimeo.com/imaginaldisc/nme01

Charlie Rose Brain Series 2 on consciousness: http://www.charlierose.com/view/interview/12025

"Soul Has Weight, Physician Thinks": *New York Times* archive, March 11, 1907 http://query.nytimes.com/mem/archive-free/pdf?res=9D07E5DC1 23EE033A25752C1A9659C946697D6CF

Ruthven, Alexander. "Kevin Bacon Is the Center of the Universe," April 7, 1994. Retrieved June 30, 2012. https://groups.google.com/forum/?from groups#!topic/rec.arts.movies/-qNue6RwTn8

The Oracle of Bacon: http://oracleofbacon.org

Eric McClean's final video: http://youtu.be/XL8PePogHbQ

2002 USA commercial: http://www.youtube.com/watch?v=afJRKbBEr2Q

9. The Accidental Fabulist

Addis, D. R., and Schacter, D. L. (2012) "The Hippocampus and Imagining the Future: Where Do We Stand?" *Frontiers in Human Neuroscience 5*: 173.

Addis, D. R., et al. (2007) "Remembering the Past and Imagining the Future: Common and Distinct Neural Substrates During Event Construction and Elaboration," *Neurophysologia* 45: 1363–377.

Ball, C. T., and J. Hennessey. (2009) "Subliminal Priming of Autobiographical Memories," *Memory* 17: 311–22.

Boyd, Brian. *On the Origin of Stories: Evolution, Cognition and Fiction.* Cambridge: Belknap/Harvard University Press, 2009.

Carey, Benedict. "Decoding the Brain's Cacophony," *The New York Times,* October 31, 2011.

———. "This Is Your Life and How You Tell It," *The New York Times,* May 22, 2007.

Carry, M. et al. (1996) "Imagination Inflation: Imagining a Childhood Event Inflates Confidence That It Occurred," *Psychonomic Bulletin and Review* 3: 208–14.

Conway, M. A., and C. W. Pleydell-Pearce. (2000) "The Construction of Autobiographical Memories in the Self-Memory System," *Psychological Review* 107: 261–88.

Deese, J. (1959) "On the Prediction of Occurrence of Particular Verbal Intrusions in Immediate Recall," *Journal of Experimental Psychology* 58: 17–22.

Fan, Jiayang. "Mike Daisey's Pride," *The New Yorker,* March 17, 2012.

Fivush, Robyn, and Catherine Haden, editors. *Autobiographical Memory and the Construction of a Narrative Self.* Mahwah, NJ: Erlbaum, 2003.

Forster, E. M. *Aspects of the Novel.* New York: Mariner Books, 1956.

Gazzaniga, Michael S. *Who's in Charge? Free Will and the Science of the Brain.* New York: HarperCollins, 2011.

———. "The 'Interpreter' in Your Head Spins Stories to Make Sense of the World," *Discover,* August 1, 2012.

Goldfield, Hannah. "The Art of Fact-Checking," *The New Yorker,* February 9, 2012.

Gottschall, Jonathan. *The Storytelling Animal: How Stories Make Us Human.* New York: Houghton Mifflin, 2012.

Habermas, T., and S. Bluck. (2000) "Getting a Life: The Emergence of the Life Story in Adolescence," *Psychological Bulletin* 126: 748–69.

Hamzelou, Jessica. "The Manhattan Memory Project," *New Scientist*, September 10, 2011.

Hassabis, D. et al. (2013) "Imagine All the People: How the Brain Creates and Uses Personality Models to Predict Behavior," *Cerebral Cortex*, March 5, 2013.

Hasson, U. et al. (2004) "Intersubject Synchronization of Cortical Activity During Natural Vision," *Science* 303: 1634–640.

Hayden, Erica Check. "Internal Monologue: Mike Daisey and the Predictability of Lies," *The Last Word on Nothing,* March 21, 2012.

Hood, Bruce. *The Self Illusion: How the Social Brain Creates Identity.* Oxford, UK: Oxford University Press, 2012.

Howard, G. S. (1991) "Culture Tales: A Narrative Approach to Thinking, Cross-Cultural Psychology, and Psychotherapy," *American Psychologist* 46: 187–97.

Hsu, Jeremy. "The Secrets of Storytelling: Why We Love a Good Yarn," *Scientific American*, September 18, 2008.

Inception. (2010) Warner Bros. Written and directed by Christopher Nolan.

Inglis-Arkel, Esther. "A Couple of Projects Show How Easy It Is to Create Fake Memories," io9.com, January 20, 2012.

Kelly, Maura. "Why Storytellers Lie," *Atlantic*, April 2012.

Konnikova, Maria. "Hunters of Myths: Why Our Brains Love Origins," *Literally Psyched*, *Scientific American*, April 7, 2012.

———. "Our Storytelling Minds: Do We Ever Really Know What's Going on Inside?" *Literally Psyched*, *Scientific American*, March 8, 2012.

Kornell, Nate. "The Fatal Flaw of the Storyteller," *Everybody Is Stupid Except You*, *Psychology Today*, March 27, 2012.

Lacey, S. et al. (2012) "Metaphorically Feeling: Comprehending Textural Metaphors Activates Somatosensory Cortex," *Brain and Language* 120: 416–21.

LeDoux, J. E. et al. (1977) "A Divided Mind: Observations on the Conscious Properties of the Separated Hemispheres," *Annals of Neurology* 2: 417–21.

Lee, S. S., and M. Dapretto (2006) "Metaphorical vs. Literal Word Meanings: fMRI Evidence Against a Selective Role of the Right Hemisphere," *Neuroimage* 15, 29: 536–44.

Lehrer, Jonah. "Head Case: When Memory Commits an Injustice," *Wall Street Journal*, April 13, 2012.

Lewis, C. S. *Till We Have Faces: A Myth Retold*. New York: Harcourt Brace & Company, 1980.

Lieblich, A., D. McAdams, and R. Josselson, editors. *Healing Plots: The Narrative Basis of Psychotherapy*. Washington, D.C.: American Psychological Association, 2004.

Loftus, Elizabeth. "Creating False Memories," *Scientific American*, September 1997.

Loftus, Elizabeth, and Katherine Ketcham. *The Myth of Repressed Memory*. New York: St. Martin's Press, 1994.

Lombrozo, Tania. (2011) "The Instrumental Value of Explanations," *Philosophy Compass* 6: 539–51.

———. (2006) "The Structure and Function of Explanations," *Trends in Cognitive Sciences* 10: 464–70.

Maguire, E.A., and D. Hassabis. (2011) "Role of the Hippocampus in Imagination and Future Thinking," *Proceedings of the National Academy of Sciences* 108: E39.

Mar, R. A. (2011) "The Neural Bases of Social Cognition and Story Comprehension," *Annual Review of Psychology* 62: 103–34.

Mar, R.A., et al. (2006) "Bookworms Versus Nerds: The Social Abilities of Fiction and Non-Fiction Readers," *Journal of Research in Personality* 40: 694–12.

———. (2009) "Exploring the Link Between Reading Fiction and Empathy: Ruling Out Individual Differences and Examining Outcomes," *Communications: The European Journal of Communication* 34: 407–28.

Marshall, Jessica. "Future Recall," *New Scientist*, March 24, 2007.

McAdams, D. P. (1996) "Personality, Modernity, and the Storied Self: a Contemporary Framework for Studying Persons," *Psychological Inquiry* 7: 295–321.

———. (2006) "The Redemptive Self: Generativity and the Stories Americans Live By," *Research in Human Development* 3: 81–100.

McAdams, D. P., and J. M. Adler. "Autobiographical Memory and the Construction of a Narrative Identity: Theory, Research and Clinical Implications." In *Social Psychological Foundations of Clinical Psychology*. Edited by James E. Maddux and June Price Tangney. New York: The Guilford Press, 2010, 36–50.

McAdams, D. P., and J. L. Pals. (2006) "A New Big Five: Fundamental Principles for an Integrative Science of Personality," *American Psychologist* 61: 204–17.

McNerney, Sam. "What Popular Psychology Books Forget: The Danger of Storytelling," *Why We Reason,* January 7, 2012.

Nisbett, R., and T. Wilson. (1977) "Telling More than We Can Know: Verbal Reports on Mental Processes," *Psychological Review* 84: 231–59.

Ouellette, Jennifer. "Meet Me Halfway," *Cocktail Party Physics, Scientific American,* January 31, 2012.

———. "Narnia, Schmarnia," *3 Quarks Daily,* May 8, 2006.

———. "What's Your Story? The Psychological Power of Narrative," *Nautilus,* May 1, 2013.

Patel, Ushma. "Hasson Brings Real Life into the Lab to Examine Cognitive Processing," *Princeton University,* December 5, 2011.

Paul, Annie Murphy. "Your Brain on Fiction," *The New York Times,* March 17, 2012.

Pennebaker, J. W., and J. D. Seagal. (1999) "Forming a Story: The Health Benefits of Narrative," *Journal of Clinical Psychology* 55: 1243–254.

Peralta, Eyder. "This American Life Retracts Mike Daisey's Apple Factory Story," *The Two Way,* NPR.org, March 16, 2012.

Roediger, H. L., and K. B. McDermott. (1995) "Creating False Memories: Remembering Words not Presented in Lists," *Journal of Experimental Psychology: Learning, Memory, and Cognition* 21: 803–814.

Roediger, H. L. et al. "The Role of Associative Processes in Producing False Remembering." In *Theories of Memory II.* Edited by Martin A. Conway, Susan Gathercole, and Cesare Cornoldi. Hove, Sussex: Psychological Press, 1998, 187–245.

Roser, M., and M. S. Gazzaniga. (2004) "Automatic Brains—Interpretive Minds," *Current Directions in Psychological Science* 13: 56–9.

Rubin, David C., editor. *Remembering Our Past: Studies in Autobiographical Memory.* Cambridge, MA: Cambridge University Press, 1996.

Schacter, Daniel. *The Seven Sins of Memory: How the Mind Forgets and Remembers.* New York: Houghton Mifflin, 2001.

Schacter, D. L., and D. R. Addis. (2009) "Remembering the Past to Imagine the Future: A Cognitive Neuroscience Perspective," *Military Psychology* 21: S108–S112.

Schacter, D. L. et al. (2011) "Memory Distortion: An Adaptive Perspective," *Trends in Cognitive Sciences* 15: 467–74.

Singer, J. A. (2004) "Narrative Identity and Meaning Making Across the Adult Lifespan: An Introduction," *Journal of Personality* 72: 437–59.

Spreng, R. N., and C. Grady. (2010) "Patterns of Brain Activity Supporting Autobiographical Memory, Prospection and Theory-of-Mind and Their Relationship to the Default Mode Network," *Journal of Cognitive Neuroscience* 22: 1112–123.

Squire, L. R. et al. "Role of the Hippocampus in Remembering the Past and Imagining the Future," *Proceedings of the National Academy of Sciences* 107: 19044–9048.

Stephens, G. J. et al. (2010) "Speaker-Listener Neural Coupling Underlies Successful Communication," *Proceedings of the National Academy of Sciences* 107: 14425–4430.

Szpunar, K. K., and K. B. McDermott. (2008) "Episodic Future Thought and its Relation to Remembering: Evidence from Ratings of Subjective Experience," *Consciousness and Cognition* 17: 330–34.

Szpunar, K. K. et al. (2007) "Neural Substrates of Envisioning the Future," *Proceedings of the National Academy of Sciences* 104: 642–47.

Tulving, E. (2002) "Episodic Memory: From Mind to Brain," *Annual Review of Psychology* 53: 1–25.

Weinstein, Y. et al. (2010) "True and False Memories in the DRM Paradigm on a Forced Choice Test," *Memory* 18: 375–84.

Wolman, David. "The Split Brain: A Tale of Two Halves," Nature.com, March 14, 2012.

Online Resources

The Redemptive Self (online): http://www.redemptiveself.northwestern.edu/toc

"Mr. Daisey and the Apple Factory" (retracted; original air date: January 6, 2012)   http://www.thisamericanlife.org/radio-archives/episode/454/mr-daisey-and-the-apple-factory

NPR Retraction, March 16, 2012: http://www.thisamericanlife.org/radio-archives-episode/460/retraction

The False Memory Archive: http://falsememoryarchive.com

INDEX

Black Bodies and Quantum Cats
Tales from the Annals of Physics
Foreword by Alan Chodos

A delightful read for armchair physicists, this book traces the history of the field and draws examples from popular culture to show that physics is not an arcane pursuit but rather a fundamental part of our everyday world.

ISBN 978-0-14-303603-6

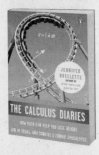

The Calculus Diaries
How Math Can Help You Lose Weight, Win in Vegas, and Survive a Zombie Apocalypse

Ouellette confronts her math phobia and learns to apply calculus to everything from gas mileage to dieting, proving that even the mathematically challenged can master the universal language.

ISBN 978-0-14-311737-7

The Physics of the Buffyverse

In the tradition of *The Physics of* Star Trek, Ouellette explains concepts in the physical sciences through examples culled from the weird and wonderful worlds of the hit TV shows *Buffy the Vampire Slayer* and its spin-off, *Angel*.

ISBN 978-0-14-303862-7

PENGUIN BOOKS